Peter Sprent

Statistics
in Action

PENGUIN BOOKS

Penguin Books Ltd,
Harmondsworth, Middlesex, England
Penguin Books,
625 Madison Avenue, New York, New York 10022, U.S.A.
Penguin Books Australia Ltd,
Ringwood, Victoria, Australia
Penguin Books Canada Ltd,
2801 John Street, Markham, Ontario, Canada L3R 1B4
Penguin Books (N.Z.) Ltd,
182–190 Wairau Road, Auckland 10, New Zealand

First published 1977
Reprinted 1979

Made and printed in Great Britain by
Richard Clay (The Chaucer Press) Ltd,
Bungay, Suffolk
Set in Monotype Times

To Janet –
Many happy landings in L A

Contents

Preface

This is a book about the way statisticians look at data. Their approach is illustrated by examples.

There is a constant flow of new ideas in applied statistics; the validity of established techniques is often questioned. This is a healthy sign in a developing subject, but it does make things confusing for the user. One of my aims is to reduce this confusion by pointing out the strengths and weaknesses of various concepts.

Some of the techniques described cannot be unreservedly recommended. For example, I discuss the much too widely used concepts of hypothesis testing and of correlation in some detail largely to make clear their limitations.

While the examples reflect my personal interests they have been selected from a number of fields of application. Some of the topics are well established. Others are relatively new. Some are more fashionable than others.

The book was inspired by a short course of lectures given in Dundee in 1974 to staff and postgraduate students from many disciplines. I hope it will interest general readers who want to get the 'feel' of statistics, or be a prelude to further study and a complement to formal elementary textbooks for those who wish to go into the subject in more detail. Recommended further reading is to be found in the Bibliography.

I am indebted to the Society of Experimental Psychology and to Gina Geffen and her co-authors for permission to quote the data in Table 9 from a paper in the *Quarterly Journal of Experimental Psychology* and to Frank Yates and the publishers, Messrs Oliver & Boyd, for permission to reproduce abridged versions of the normal, *t*, *F* and chi-squared tables from *Statistical Tables for Biological, Agricultural and Medical Research*.

Many imperfections in earlier drafts have been eliminated as a result of helpful comments from two of my colleagues, Dick Brown and Tony O'Hagan. My wife also proved a responsive guinea-pig by giving a numerate biologist's reaction to the text and Penguin's advisory editor, David Nelson, suggested a number of valuable clarifications.

Dundee, 1976   P.S.

1 The World of Numbers

A simple beginning

The United Kingdom trade deficit in January 1975 was £275 million. On St Patrick's Day that same year £1 was worth US$ 2.42. A few days later an experimenter found that the weight gains of six pigs during one week varied between 1217 and 1314 grams. These figures, or *statistics* are all of interest or importance to some people.

These statistics are numbers that mean something, but it's quite a business to know what some other numbers mean, if indeed they mean very much at all. Opposite my window there's a bill poster saying:

DON'T RISK IT
USE A CROSSING – IT'S UP TO FIVE TIMES SAFER.

That's all it says. Of course this poster carries a message; it's telling me I'd be wise to use a marked pedestrian crossing to get to the other side of the road. Fine, but what does 'up to five times safer' mean?

It certainly doesn't mean I'd be up to five times safer if I spent the whole day walking back and forth over a pedestrian crossing instead of writing a book. It only means that in some sense I shall be much safer using a crossing than I would be if I just went over anywhere; but even that statement needs qualification. Pedestrian crossings are usually at busy points; if I can cross the road where traffic is light it is unlikely to be safer to walk several hundred yards to a busy area *just* to use a crossing. If I use a handy crossing rather than just barge out willy-nilly into heavy traffic I should be safer. But how is the 'five times' arrived at? Have accident rates been compared for people using crossings with those for people who cross within, say, 200 metres of them?

We are not told, and indeed we needn't worry very much so long as the poster encourages people to use a crossing when one is available. So the 'five times' really isn't very important.

When numbers are more important we can't be so casual about their meaning. A gloomy set of monthly trade figures like those for January 1975 may knock many millions of pounds off share prices, just as a dramatic rise in the cost of living index may trigger off wage claims with wide economic repercussions.

This book is about getting as much relevant information as we can from certain numbers. The numbers we shall be talking about will often be *sets* of data like weights of pigs or daily counts of defective items turned out by a machine rather than *single* statistics such as a money exchange rate or a cost of living index.

In everyday use the word 'statistics' has two meanings. It refers firstly to the numbers, or data, themselves and secondly to the science or craft of abstracting useful information from sets of numbers.

Statistics in the first sense are all-important to governments and to industry. Actions are clearly influenced by the vast array of statistics on trade, finance, taxation, imports, exports, earnings, employment and other matters readily available from official sources. There are examples in the Penguin *Facts in Focus* (1974) and many more in the U.K. *Monthly Digest of Statistics* and other publications from the Central Statistical Office and its counterparts in other countries.

One of the worrying things about figures is the way different groups – for example opposing political parties – can put very different interpretations on the same figures. Interpretation of data is a skilled business. Many of the pitfalls are amusingly described in Darrell Huff's (1973) *How to Lie with Statistics*. In more serious vein Bartholomew and Bassett (1971) look at the quantitative approach to human affairs.

Figures all too often become meaningless when divorced from their context. That trade deficit of £275 million for January 1975 was referred to by the *Financial Times* as 'freakishly good'. That may have been fair comment in the 1975 economic climate, but forty years ago would have spelled economic disaster.

The task we have set ourselves in this book involves *statistics* in both senses of the word, but more especially the second. Much of the data we use in our examples have been obtained from small surveys or experiments and we shall look at how they may be used to draw conclusions of wide applicability.

At the start we referred to weekly weight gains of six pigs varying between 1217 and 1314 grams. These data reflect a common phenomenon; even if pigs as alike to one another as we can get them are treated in the same way they will not all gain exactly the same weight in a week.

Further, we could be pretty certain that another group of pigs in the same week or the same group in a different week would have different weight gains.

An important statistical problem is to disentangle such natural and uncontrollable variation between similarly treated individuals or units (in this example pigs) from differences that might arise from known and monitored changes in circumstances. For example we might feed one group of six pigs on a certain diet and another group of six on a different diet, keeping all other conditions the same. We should like to know if diet affects weight gain. If one diet gives the range of weekly gains from 1217 to 1314 grams and the other gives a range from 893 to 947 grams it is clear which is the better diet. If the second diet gives gains between 1198 and 1283 grams we need a statistician to tell us whether we can really conclude that either diet is superior.

Facts behind figures

Data referring to a limited number of experimental units, called in statistical jargon the *sample*, can be used to answer questions about the larger collection of units, called the *population*, from which the sample is drawn.

There are many statistical recipe books, including M. J. Moroney's (1965) *Facts from Figures*, that give excellent accounts of how to do calculations with specific formulae in particular situations. Useful as recipe books are, it is helpful not only to students of statistics but to experimenters, industrialists, sociologists – indeed to anyone who collects data – to know a little about the approach and attitude of the statisticians who draw up the recipes. We shall show why certain techniques are used and what they can achieve, leaving out unnecessary technical details; we shall look at the limitations and restrictions of various approaches in a treatment which is not exhaustive – and we hope not too exhausting.

Some techniques we shall meet are of restricted use only. Two examples are *significance testing* and the calculation of *correlation coefficients*; we shall look at them in detail because they are used in practice widely – in fact more widely than they should be. We describe them fully to expose their weaknesses and to show why other concepts should often be used in their place.

Later chapters are each devoted to a particular aspect of the statistician's craft – a term we choose because his profession is a strange mixture of art and science. He must employ art in devising a mathematical model that reflects the mechanisms giving rise to observations. He must employ science within the framework of strict mathematical logic to make deductions.

The mathematical basis of statistics gives it great power because one mathematical model may be valid for data in a number of fields ranging perhaps from political science through industry to medicine or agriculture. In Chapter 3 we shall meet a mathematical model applicable to politics, medicine and commerce. In each case the data take the form of counts of items (people or goods) with various characteristics.

Not all data consist of counts; often they are measurements. Weight gains of pigs subjected to various diets are examples of measurements; in this problem we want to know whether one diet will *on average* give a higher weight gain per week than another diet.

The complicating factor is the natural variation between pigs even when they start off as alike as we can get them and are all treated the same way. Such natural variation crops up all over the place in statistics – indeed it is part of the *raison d'être* of the subject – and masquerades under a variety of names; *error variation* is a common one but for biological material at any rate it is not particularly appropriate. Because Joe is taller than Bill one cannot say that the height of either is in error. A more modern term for uncontrolled natural variation is *noise*, a term carried over to statistics from communication theory and familiar to us as static or background noise. We shall look at the comparison of different experimental conditions in the presence of noise in Chapters 4 and 5.

In talking about pigs in our first example we are following history, for it was in agricultural experiments that many statistical techniques had their beginnings. Although it uses many long-established mathematical concepts, particularly the idea of probability, statistics is really a twentieth-century subject. R. A. Fisher (1890–1962), the father of modern methods in experimental design and analysis, carried out his pioneering work with agricultural experiments at Rothamsted Experimental Station in Hertfordshire. Many of the techniques developed there are now used with little modification in such diverse fields of application as industry, medicine and the social sciences. Statistics has even found a place in 'detective' work in linguistic studies, a facet we illustrate in Chapter 6.

Coming to grips with variation

The error variation or noise that expressed itself by weight differences in the pig feeding example has an unpredictable element for an individual. Fed in the same way, some pigs do a little better than average, others a little worse. Sometimes we can reduce such unwanted variation by careful control of experimental conditions, selecting pigs as alike as possible, and keeping them in a similar environment during the experiment. Such precautions may be costly and they will not entirely eliminate 'noise' for they cannot overcome differences attributable to heredity.

'Noise' implies that using another set of pigs in our experiment would give different results. The variation induced by noise is referred to in this context as *random sampling variation*, or more loosely simply as *random variation* or as *sampling variation*. 'Random' describes its unpredictable nature and 'sampling' indicates that different samples or experimental units (in our example pigs) would give different results.

In this experiment we are essentially interested in deciding whether *on average* one diet is superior and if so by how much. We obtain precise figures on weight increase only for the particular *sample* of pigs in our experiment, but statistics provides tools for doing something more important and interesting: to infer things about the results if we applied one of the diets to the *population* of all pigs similar to the ones used in the experiment.

In this situation we are interested in *systematic differences* induced by diets, but the presence of noise makes it difficult to determine the true value of the systematic differences.

Sorting out systematic differences and noise or random variation is only part of a statistician's job. Many systems have a short-term behaviour that has a basically unpredictable or random element, but from which there emerges a long-term regularity or predictability. Systems with an essential random element in their pattern are called *stochastic processes*, a grandiloquent name for something that proceeds randomly.

A large post office is a useful setting to study such a process, that of queueing. Typically a number of counter positions or serving points provide for a wide range of demands from customers; at one extreme an elderly lady wants just one postage stamp (second class) and at the other an office clerk wants many pounds-worth of postage and insurance stamps of differing denominations. There are individuals wanting postal orders, collecting pensions, paying television or car licences and others enquiring about overseas postal rates or buying savings bonds.

From past experience the post office can predict with a fair degree of accuracy the number of people likely to patronize them on a given day. They also have a good idea of the average time needed to serve a customer and in very broad terms how this varies from customer to customer. There is still a large degree of uncertainty about when the next customer will arrive and how long a particular customer will take to serve.

The system is further complicated by the unpredictable behaviour of an individual who enters the post office. Will he or she join the shortest queue? The queue nearest the door? Or the queue that is being served by the pretty blonde postal clerk? Will the customer perhaps select a queue by trying to make some estimate, or 'guesstimate', of the time it is likely to take to serve people already waiting? If he sees in one queue a lot of people holding car licence renewal forms he may avoid it because he knows the transaction is a lengthy one. He might also avoid a queue where there are several clerks holding what look like long lists of requirements. A smart customer who can make a rapid assessment of how long each queue will take to evaporate joins the queue – not necessarily the shortest – which he believes will be served most rapidly. The system is further complicated by people changing queues if they think another is moving more rapidly than the one they are currently in.

In the interests of good customer relations the post office does not want to keep customers waiting; it must provide an adequate number of service points. At the same time it wants to keep its salary bill low and does not want to employ counter clerks who are idle for long periods.

Reorganization of a system can sometimes reduce running costs and at the same time produce better service by eliminating bottlenecks in the system. Some years ago the larger British post offices had separate counters for each type of business – one selling stamps, one issuing licences, one handling postal orders and so on. It was not uncommon to find no customers at a stamp counter, say, while a long queue waited to renew television licences. In recent years most post offices have changed to multi-purpose counters so that a customer wishing to buy some stamps and renew his television licence may do so at the one counter. The change seems to have pleased both customer and post office, and average waits have been reduced.

Travellers will be familiar with the wide variety of queueing systems in banks; in some countries one may have to queue at two or more counters to complete a withdrawal – there are marked differences between the systems in the U.K., the U.S.A. and various Continental countries.

In organizing a queueing system one tries to make its behaviour steady, in the sense that seldom will there be large queues nor will servers be idle for long. Even a steady system can easily be upset by small changes in conditions. A system with three servers may work well when customers arrive at an average of one per minute, but if the average interval between arrivals drops to fifty-seven seconds the system may be thrown into chaos unless service time is speeded up, queues simply getting longer and longer.

Queueing systems are not confined to people; they may apply to arrivals of ships at a port, to trucks delivering or collecting at a factory, or to the build-up and dispersal of items from a store. There are closely related problems in studying the behaviour of water-storage dams and reservoirs with changing patterns of supply and demand.

The statistician has built up an appreciable body of theory for studying queueing systems and other stochastic processes. In addition computers have enabled him to set up *simulation* models to reproduce in a matter of seconds patterns of behaviour of, for example, a queueing system extending over a period of weeks or even years. Simulations can also examine the effect upon the system of such changes as putting on additional servers or speeding up service times. We take a closer look at all this in Chapter 8.

More about variation

We've talked about variation in pigs and post office queues. Let's think about it – and what we want to do about it – in some other situations.

Where random variation is simply *noise* we may want to eliminate it. In finding out which variety of wheat does best in certain climatic and soil conditions we would try to stabilize these factors as far as possible in any experiment we did. We cannot eliminate all noise and much statistical effort is devoted to measuring it and disentangling it from differences we want to measure.

Random variation is with us all the time. Manufacturers strive for uniform products and want to reduce 'noise' in the form of quality variation. They succeed to varying degrees, sometimes reflected in prices of products. Success is seldom complete. Not all sparking plugs of a particular type have exactly the same useful life. The specific gravity of our favourite beer varies slightly from brew to brew. A weighing machine set to deliver 500 grams of sugar into a bag will not deliver exactly 500 grams each time; this is why labels often state *minimum* contents or *average* contents.

If nuisance variation is trivial or unimportant it may not be worth trying to eliminate. Also, variation that is simply nuisance variation in one situation may be important in another. If, for example, patients react differently to a drug it may be important to track down why. Whether or not the reaction is influenced by a patient's sex, age, diet, blood group, blood pressure, occupation or consumption of alcohol concern a clinician prescribing the drug for a particular patient. These same factors may represent sources of nuisance variation in an experiment to determine the overall effectiveness of one drug relative to another. This example gives another instance of a situation where we might control some but not all nuisance variation. A clinician may be able to adjust a patient's blood pressure by suitable treatment, but he cannot alter the patient's sex, age or blood group.

Social surveys are another area where nuisance variation in one context may be the subject of study in another. National, religious or ideological differences often fall into this category. If we want to know whether Englishmen, Frenchmen, Germans and Chinese have different views on the importance of having a bathroom in one's house the differences between national attitudes are of interest. If, on the other hand, we want to take a sample of people in a town to estimate how many of the population have bathrooms in their homes, then any mixture of nationalities in the city with differing attitudes towards the need for bathrooms may well introduce sampling variation.

Similarly, in a study of the attitudes of different religious groups towards divorce, differing attitudes of individuals *within* any one group represent nuisance variation when one is looking at basic differences between groups. In other studies differences in attitudes of people holding similar religious beliefs may be of interest in their own right.

In many experimental situations we *induce* variation deliberately by choice of material or application of treatments. The

experimenter then wants to measure the effect of induced variation against a background of noise.

Before the statistician can help the experimenter with this task
he may have important incidental jobs such as summarizing the
data before their message can be digested; but in this book we
shall be more concerned with deciding how much and what type
of data are needed to give informative answers to questions, and
how we go about getting those answers.

Conservation of resources

To know how much and what type of data need to be collected to
answer a particular question calls for a joint effort by the experimenter and the statistician, for statistics is truly a multidisciplin-
ary subject.

In an era when we are becoming ever more aware of the need
to conserve resources – raw material or money – we learn to
appreciate the value of collecting data. Not only in scientific
experiments, but in industry and commerce as well, collecting and
analysing data may be the only way to answer important questions. The right answer to a technological problem may save a
large company thousands of pounds or dollars or conserve
dwindling stocks of a commodity. Experiments must be planned
to get the right information as quickly as possible – often in very
complex situations.

For example, a chemical plant producing an organic compound
may give very different yields if one or more of a host of factors
such as purity of reactants, temperature, pressure, humidity, time
of processing, etc., are altered. These factors tend to interact with
one another and it can be very much a trial-and-error game to find
the best combination. The statistician can minimize the hit-or-
miss element by channelling the effort so as to get an appropriate
answer quickly.

To outline how a statistician might help with a problem like
this let us suppose yield is determined by two factors only, say
temperature and pressure. In an experiment we measure yield

at various temperatures and pressures. Our aim is to find the combination of temperature and pressure that will maximize yield.

An analogy is useful to explain the way the problem can be tackled. Suppose we have a contour map from which we can determine the height, or altitude, of any point given its latitude and longitude; let us now think of temperature as equivalent to latitude, pressure as equivalent to longitude, and yield as equivalent to altitude. What we seek is the highest peak on the map, corresponding to maximum yield.

Now suppose that we are not shown the map, but we are allowed to select four points specified by latitude and longitude. We are then given the altitudes of each of the four points. From these altitudes we can form some idea of the slope of the land on the map. In seeking a peak we would get additional information by choosing four more points in a direction that appeared to be 'up hill' from our first four points. This process can be formalized mathematically and is known as the *method of steepest ascent*.

Using this analogy the experimenter chooses four temperature and pressure combinations (corresponding to latitudes and longitudes), and determines yield at each of these experimentally. He then chooses four more points in a 'direction' indicating the greatest increase, continuing until he reaches either a peak or a plateau.

The whole process is reminiscent of the way some of us climb mountains. Unless there are obvious obstacles in the way we choose the path that keeps going upwards. The two differences in the experimental situation are that we do not have a continuous picture but only know heights at selected points (corresponding to our experimental points); secondly if we repeat the experiment with the same temperature and pressure we shall usually not get the same yield because of 'noise' or sampling variation. This would be analogous to reading latitude and longitude correctly from a map but making small random mistakes about the altitude.

Duckworth (1968) describes this and other techniques known under the generic name of *response surface designs*.

Measuring several characteristics

What can we learn from measurements of two or more characteristics on the same individual? Could they help us to distinguish
between individuals of different kinds, for example? Now most of
us can tell a Chinaman from a Nigerian at a glance, but we might
be less able to distinguish between a Tanzanian and a Nigerian
unless we knew quite a lot about the two countries.

People of different nationalities exhibit not only varying quali-
tative characteristics such as skin colour, hair type and mode of
dress, but sometimes also show variations in general shape and
size that are best described numerically. Animals and different
plant cultivars also show characteristic geometrical differences.
Superimposed upon what we might conveniently term differences
at 'breed' level are the differences between individuals within a
breed.

We might want to find out how many distinct races are represented in a group of individuals. Another problem takes this
form: given the measurements of a number of individuals of two
or more races, how can we tell the race of other individuals from
their measurements? The first type of problem is known as a
classification problem and the second as a *discrimination* problem.

Here's a simple example of a discrimination problem. An
anthropologist may have a number of specimens of thigh bones
known to belong to a species of ape and another set of specimens
of the corresponding bone in primitive man. He measures the
length and breadth of each bone and plots these on a graph as in
Figure 1.

We see from the graph that there is a distinct overlap in the
lengths of the bones for the two species and also quite an overlap
in the breadths. However, when we look at the points represented
by the symbols,' · ' for men and ' × ' for apes, we see that they fall
into two groups with little overlap. On the graph the line joining
the point 40 on the length axis to the point 5 on the breadth axis
gives a very sensible boundary between the two groups, about as
good as any we could form with a straight line. Apart from a few
exceptions, thigh bones of men are all above the boundary and
those of apes are below.

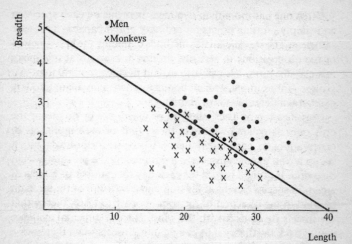

Figure 1 Lengths and breadths of thigh bones in men and monkeys

For any point exactly on the line we find that 'length + 8 × breadth' is exactly equal to 40. For points above the line 'length + 8 × breadth' exceeds 40, while for points below the line it is less than 40.

Because men are generally represented by points above the line and apes by points below, we can classify other bones of unknown origin: from men if 'length + 8 × breadth' exceeds 40, from apes if it is less than 40. We here assume all bones come either from men or apes.

In doing this we shall sometimes make a mistake, but we hope not very often. The statistician has formal ways of determining the 'best' line of the kind we fitted by eye in Figure 1. In this context we mean 'best' in the sense that the expected number of misclassifications will be kept as small as possible. We can also estimate the degree of correctness we are likely to achieve.

The function

$$\text{length} + 8 \times \text{breadth}$$

is known as a *discriminant function*, because it is on the basis of its value that we can discriminate between men and apes.

Often one has more than two measurements on each specimen and simply plotting points on ordinary graph paper is no longer sufficient. The statistician has methods of finding the best function of the observations to separate groups in a way that minimizes mis-classification. We can also extend the idea to more than two groups – for example, ancient bone specimens may belong to any of several different kinds of animal.

Classification or discrimination problems are by no means the only ones associated with two, three, four or even more sets of observations on each individual. A biologist, for instance, might set up a trap to catch insects in a path facing the prevailing wind direction. He might record for each day the number of insects of a certain species caught in his trap and a measure of the strength of wind in the direction of the trap in order to test a theory that the insects fly or drift with the wind. His experimental evidence would back the theory if he caught more insects when there was a high wind in the appropriate direction, and less when there was no wind or it blew strongly from some other direction.

We can plot our observations on a sheet of graph paper using one axis to represent the strength of wind, and the other axis to indicate the daily insect catch.

Suppose the plotted points gave a pattern something like that in Figure 2. The experimental evidence of high catches in high winds and low catches in low winds supports the theory that insects fly or drift with the wind.

The two most commonly used measures of statistical support in studying relationships between variables, or association between variables, are *correlation* and *regression*. Correlation is all too often used and we say some harsh words about it in Chapter 7, where we also discuss regression, a much more useful concept which tells us a great deal about the nature of relationships between variables.

One note of caution is needed. Relationships between variables, especially more or less straight-line relationships, are often taken to imply cause and effect. In the case of wind strength and insect catch it is perfectly reasonable to suggest that wind strength influences catch. This is a natural explanation that depends upon our prior concepts of how nature works. We would not deduce

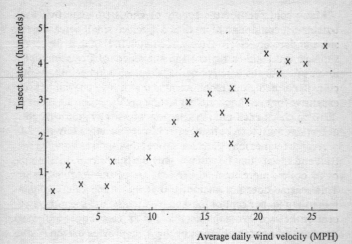

Figure 2 Insect catches in varying winds

from our experimental results that the number of insects caught influenced the wind strength! This would fly in the teeth of common sense, and be completely at variance with our knowledge of physical and biological principles. We have lots of information about the causes of wind and their dynamic effects upon insects.

On the other hand, if we discover a strong straight-line relationship between the number of ice-creams sold in any year and sales of sunglasses we need not seek explanations of cause and effect; both figures are influenced by climate and provide no evidence that consuming ice-cream alters the sensitivity of one's eyes to light, or that wearing sunglasses induces a craving for ice-cream.

The commercial angle

Increasing use is made of statistical ideas in business and commerce. Large retailers are very keen to ensure that the products they sell in different branches are comparable in quality and also that quality does not change dramatically from week to week.

Many contracts for the supply of goods stipulate that each batch must contain no more than a specified small percentage of sub-standard or defective items. If a batch fails in this respect the supplier must either replace the sub-standard items, accept a lower price for the batch or perhaps replace it altogether. A difficulty in enforcing this requirement is that it may be prohibitively expensive (or even impossible if testing involves destruction of an item) to check each unit. A decision whether or not a batch is acceptable must then be made after inspecting only a sample.

A large retailer may purchase several thousand crates of grapefruit and specify that he will pay only the full price if, in a sample of 20 crates, more than 98 per cent are of marketable quality. This scheme does not guarantee that the whole batch contains more than 98 per cent of marketable quality just because 98 per cent of the sample are marketable, but at least it does indicate a low level of poor fruit. Tests of this type provide examples of a technique known as *quality control*. In Chapter 9 we shall look at several methods of quality control including one modern development known as *cumulative sum charts* – usually abbreviated to CUSUM charts.

Even accountants – people who are very particular about figures – use statistical methods to save money. Here's an example. Suppose a firm receives 5123 invoices for the supply of sundry components, each invoice involving a sum of less than £5. To check each invoice may require comparison with a store delivery note; to trace the correct note and check it may take on average two minutes. Thus about 170 man-hours would be required to check all 5123 invoices; including overheads the cost would run into hundreds of pounds. Unless many fairly major discrepancies to the firm's advantage were disclosed the check would hardly justify itself. An alternative would be to take a sample of, say, 50 invoices, and check these at something like one hundredth of the cost of a full check; if the sample were found to be all correct or if only a small number of trivial errors were located, all 5123 invoices would be paid. If major errors were found in the sample it might well pay to examine all invoices. Clearly there are problems for a statistician here. One is: how large should mistakes be before a decision is made to examine all invoices?

The statistician may also be called upon to advise on how to select the sample. For example, it would not be appropriate to take a sequence of 50 invoices in the order in which the firm had received them, for they might be an atypical set, perhaps those prepared by an inexperienced clerk doing the job while somebody more competent was sick.

Ideally we should take a *random sample*, in which each invoice has an equal chance of inclusion in the sample. In this book we shall meet various dodges for obtaining random, or effectively random, samples. A machine well known to British readers for generating an effectively random sample is ERNIE, the computer that decides the winning numbers for Premium Bond prizes.

Getting down to details

In this chapter we've had a quick tour of the statistical world. Lots of important ideas and applications haven't yet been mentioned. One is the impact of the computer both upon the sort of problems we can tackle and the way we actually tackle them. We'll look at this in Chapter 11. We've said little about the interplay between theory and practice; in Chapter 12 we'll talk about some relatively new theories and their practical implications.

At the end of most chapters we've included one or two posers to help you think about the main ideas we've introduced. In case you have difficulties with these the final 'lucky number' chapter – Chapter 13 – has hints on how to tackle them, as well as further examples for real enthusiasts.

Here's a poser for which we give no hints. Perhaps you might like to think up how you would carry out an experiment to try and find out if there is any statistically acceptable evidence that 13 is, as is so often claimed, an unlucky number, or whether it is perhaps a 'lucky number' as we have just suggested.

2 Coming to Grips with a Problem

Translating reality into mathematics

Let's introduce Dr Gesser who claims he can diagnose a certain disease simply by looking at a patient's tongue. Being a modest man he doesn't claim he's always right, but only that he very often is. The claim sounds trivial, but if Dr G. is proposing an alternative to a more expensive and time-consuming method, or one more distressing to a patient, we certainly want to know if there is anything in it. His method might at least be useful for a preliminary screening of patients.

There are several ways we could go about testing his claim. One might be to present Dr G. with a number of groups each consisting of four patients, with three in each group healthy and one diseased, and ask him to pick that one, telling him there is only one. To do this we need a 'pool' of patients for whom we have a prior correct diagnosis known to us but not to Dr Gesser. If, by examining their tongues, he correctly selected the diseased patient from all or a large number of groups we might correctly conclude that Gesser was no guesser. If he were seldom correct we would have little faith in his method.

This sort of experiment is simple in concept but there are many practical points to consider. Why choose groups of four? How many groups do we present to the doctor? When we get the results how do we decide whether or not he is guessing? If he is not guessing how 'good' is he at diagnosis?

Physical resources may well limit the size of the experiment. Clearly if the disease is uncommon it may be impossible to make up more than a few groups of four with exactly one diseased patient in each. Furthermore, even if there are plenty of patients available we don't want to make more examinations than are necessary to reach a rational decision about the doctor's claim.

How should we proceed? One of the simplest things we can do (although we shall see later it is not really very informative) is to set up a *hypothesis* and see whether our experimental evidence supported it. A clearly defined hypothesis would be that of the cynic, that Dr G. is guessing.

A virtue of this clear-cut hypothesis is that it can be formulated mathematically in terms of *probabilities*. On the basis of experimental evidence we shall either accept or reject the 'guessing' hypothesis. Rejection means we have reasonable grounds for preferring an alternative. In our example the alternative will be that the doctor is doing better than mere guesswork, although one can visualize the situation where the doctor was wrong so often that he would do better if he did just guess. In the discussion that follows we shall ignore this possibility and consider only the alternatives

(i) Dr Gesser really is a guesser;

or

(ii) Dr Gesser has at least some diagnostic ability.

To tackle this problem statistically we need some simple ideas about *probability*. The theory of probability is a branch of pure mathematics, but the ideas we need can all be expressed in simple intuitive terms requiring only a little statistical jargon.

Some terminology

The test itself, in this case examining the tongues of a group of four patients and trying to select the diseased one, is called a *trial*. In this particular trial there are two possible outcomes of interest, namely, *correct* diagnosis or *incorrect* diagnosis. Each outcome is termed an *event*. Probabilities may be associated with an event or a set of events. The first probabilities we consider apply *conditionally* upon some hypothesis being true. In particular we consider the hypothesis that the doctor is guessing. If he is, then for a group of four patients there is only one chance in four that he will guess correctly; in probability terms we say that under the hypothesis of guessing there is a probability of one quarter that his guess will be correct. The fraction, one quarter, is the ratio of one outcome (the correct guess) to the total number of equally likely outcomes (the four possible guesses). More formally this is often written, using *Pr* for probability,

Pr(correct guess) $= \frac{1}{4}$.

Generally, if we wish to say that an event A has a probability *p* under some given hypothesis we write

$Pr(A) = p$.

Probabilities may take any values between 0 and 1. Broadly speaking the nearer the probability of an event to 1, the more likely is that event to occur. An event that cannot happen has zero probability, and an event that is certain to happen has unit probability, that is a probability of 1. These ideas are intuitively reasonable, but while intuition is a good servant one must never let it become the master. It might be thought that if an event can happen at all then its probability must be more than zero; but in fact this is not always true. Fortunately such mathematical subtleties need not worry us at this stage; what we need to remember is that the greater the odds, in gambling terms, in favour of an event happening, the closer is its probability to one; the greater the odds against its happening the nearer is its probability to zero.

We can now return to Dr G.'s experiment. We assume that if the doctor has no diagnostic ability, or a certain specific (but to us unknown) diagnostic ability at the start of the experiment, this situation pertains throughout the experiment: his success or failure in spotting the diseased patient in one group does not affect his success or failure with another group. The outcomes of successive trials are then said to be *independent*.

The present analysis does not apply if the doctor's ability to diagnose changes for better or worse as he proceeds. Although we exclude this possibility it is easy to visualize a real-life situation where the doctor shows diagnostic ability at the start but is upset by a sequence of mistakes and becomes more erratic as he proceeds; contrariwise, early mistakes may stimulate him to be more careful later in an attempt to redeem his claim. To avoid this possibility it would be best not to tell the doctor the outcome of individual trials until the whole experiment was completed.

The addition of probabilities – mutually exclusive events

We need some more jargon. We have called examining a group of four and selecting the supposedly diseased patient a *trial*. Correct selection is called a *success* and incorrect selection a *failure*. The set of all trials carried out constitutes an *experiment*.

At a trial *success* and *failure* are examples of *disjoint* or *mutually exclusive* events: one or other, but not both, may occur at any one trial. They are also *exhaustive* events, since they represent the only possible outcomes at a trial. Not all mutually exclusive pairs of events are exhaustive; for example, the events 'Jack has brown eyes' and 'Jack has blue eyes' are mutually exclusive but not exhaustive, for Jack may have green or grey eyes.

The probability that one of a pair of mutually exclusive events occurs is obtained by *adding* the probabilities that each will occur. In symbols, if A, B represent two mutually exclusive events we may write the probabilities (*Pr*s) as

$$Pr(\text{A or B}) = Pr(\text{A}) + Pr(\text{B}).$$

The sum of these probabilities never exceeds 1. It equals 1 when the two events are exhaustive.

In our experiment if the doctor is guessing then at any one trial

$$Pr(\text{success}) = \tfrac{1}{4}, \; Pr(\text{failure}) = \tfrac{3}{4},$$

the latter value arising because there are 3 chances in 4 that a guess will select the wrong patient. The sum of these probabilities is 1, consistent with the mutually exclusive and exhaustive nature of the events; this result also conforms to the notion of an event that is certain to happen having unit probability.

The addition rule has a natural extension to three or more events when only one of these can occur at any trial, i.e. they are mutually exclusive. For example, if A, B, C, D are four mutually exclusive events then

$$Pr(A \text{ or } B \text{ or } C \text{ or } D) = Pr(A) + Pr(B) + Pr(C) + Pr(D).$$

Now let us suppose that the doctor is guessing and that he performs two trials. What is the probability that he selects the correct patient from both groups? For this dual success he must select the one correct pair from all possible pairings from the two groups. There are sixteen possible pairings, combining any one of the four in the first group with any one of the four in the second. If we call the four patients in the first group A, B, C, D and those in the second W, X, Y, Z the possible pairings are AW, AX, AY, AZ, BW, BX, BY, BZ, CW, CX, CY, CZ, DW, DX, DY, DZ. If C and X are the diseased patients the pair CX represents dual success and if Dr G. really is a guesser he has only one chance in sixteen of making the right pairing. In probability terms

$$Pr(\text{two successes}) = \tfrac{1}{16}.$$

The multiplication of probabilities – independent events

Remember we assume that the outcomes of two trials are *independent*; this implies that the probability associated with any event in one trial is not influenced by what happens at the other trial. In these circumstances we may use a simple *multiplication rule* that states that if two events are independent the probability that both will occur is the product of the probabilities of each occurring. In symbols, if A, B represent two independent events we may write

$$Pr(A \text{ and } B) = Pr(A) \times Pr(B).$$

Thus, for the doctor to achieve two successes, one at the first and the other at the second trial, we have

$$Pr(\text{two successes}) = Pr(\text{success at first}) \times Pr(\text{success at second})$$
$$= \tfrac{1}{4} \times \tfrac{1}{4}$$
$$= \tfrac{1}{16}.$$

We may use the multiplication rule to calculate other probabilities associated with two trials on the assumption of guessing. The relevant ones are

Pr(fail at both) $= Pr$(fail at first) $\times Pr$(fail at second)
$$= \tfrac{3}{4} \times \tfrac{3}{4}$$
$$= \tfrac{9}{16},$$

Pr(1st fail, 2nd success) $= Pr$(1st fail) $\times Pr$(2nd success)

$$= \tfrac{3}{4} \times \tfrac{1}{4}$$
$$= \tfrac{3}{16},$$

and

Pr(1st success, 2nd fail) $= Pr$(1st success) $\times Pr$(2nd fail)
$$= \tfrac{1}{4} \times \tfrac{3}{4}$$
$$= \tfrac{3}{16}.$$

The multiplication rule can be extended to more than two *independent* events and takes the form that the probability of *all* occurring is the product of the probabilities of each occurring. Thus for four mutually independent events A, B, C, D

$$Pr(\text{A and B and C and D}) = Pr(\text{A}) \times Pr(\text{B}) \times Pr(\text{C}) \times Pr(\text{D}).$$

For example, with three trials under the hypothesis that the doctor is guessing we find

$$Pr(\text{success at all three trials}) = \tfrac{1}{4} \times \tfrac{1}{4} \times \tfrac{1}{4} = \tfrac{1}{64}$$

and

$$Pr(\text{1st success, 2nd fail, 3rd fail}) = \tfrac{1}{4} \times \tfrac{3}{4} \times \tfrac{3}{4} = \tfrac{9}{64}.$$

To check that you have the hang of things you might like to verify that the probability of success at the first and third trials and failure at the second would be $\tfrac{3}{64}$.

The probability of $\tfrac{1}{64}$ for 3 successes without any failures implies that if the doctor had been guessing there is only 1 chance in 64 that he would be correct at all three trials. Similar arguments (try working them out for yourself) easily establish that if four trials are performed there is only 1 chance in 256 of his being right each time and that the chance of his being right in all of five trials is only 1 in 1024.

We really see the power of the multiplication rule in this last case. It would have taken a long time to write down all the 1024 'quintuplets' of possible selections of one patient from each of the five groups.

What should we think if the doctor picks the correct patient each time in an experiment consisting of 5 trials? Surely our sensible reaction is that he has established his claim? But those of us who are gamblers at heart know that long shots do come off sometimes, and we have just established that even with guessing there is this 1 in 1024 chance of being correct each time. The odds are even more remote of winning a big lottery prize, the football pools or a major Premium Bond prize – but somebody does win.

Suppose we follow our intuitive ideas though, and decide to *reject* the hypothesis that Dr G. is a guesser if he is right at all five trials. If we do this we accept a rather vague hypothesis that he can do something at least a bit better. In doing so we are still being somewhat arbitrary. Some may feel we have not performed enough trials to remove all reasonable doubt while others may feel we could have made a decision earlier – perhaps after four trials – without leaving much doubt. Let's now tie up some of the loose ends.

Conducting an experiment

There are certainly no hard and fast rules about how many trials we should perform before reaching a decision, or about precisely what evidence should lead us to accept or reject a hypothesis. What is important, and what statistics enables us to do, is to back up conclusions with probability statements that indicate the strength of the evidence leading to those conclusions. The phraseology is necessarily vague at this stage but what we mean will become clearer as we proceed.

We must set up precise rules for conducting an experiment before we start it and examine the implications of these rules for various hypotheses.

Suppose we lay down the following rules for the diagnostic experiment:

1. If Dr G. makes a mistake at or before the fourth trial the experiment immediately stops; we accept the hypothesis that he is guessing.

2. If he completes four trials without any mistakes the experiment then stops: we reject the hypothesis that he is guessing and accept the hypothesis that he has some diagnostic ability.

Before we even start the experiment we can calculate the probabilities of each of these two outcomes *on the assumption that the doctor is guessing.* Rule 1 tells us that the experiment ends at the first trial if the doctor selects the wrong patient, and we have already seen that at a single trial

$Pr(\text{failure}) = \frac{3}{4}$

The experiment ends at the second trial if and only if there is success at the first trial followed by failure at the second. The multiplication rule for probabilities (page 32) shows that this combination of events has a probability of $\frac{3}{16}$, so this is the probability of stopping after the second trial.

Let us now work out the probability of stopping at or before the completion of two trials, i.e. *either* after the first trial *or* after the second. These two possibilities are *mutually exclusive* so we may use the *addition rule* for probabilities (page 31), which tells us that the probability of one or other of the events taking place is the sum of the probabilities of each of them taking place. Thus,

$Pr(\text{stopping after 1st or 2nd trial})$
$\quad = Pr(\text{stopping after 1st}) + Pr(\text{stopping after 2nd})$
$\quad = \frac{3}{4} + \frac{3}{16}$
$\quad = \frac{15}{16}.$

So if the doctor is just guessing there is only one chance in sixteen that we shall proceed to a third trial. We can see this in two ways. Firstly, it is the opposite event to stopping after either the first or second trial and opposite events are mutually exclusive and exhaustive; thus (page 31) it has a probability of

$1 - \frac{15}{16} = \frac{1}{16}.$

Secondly, we can proceed to a third trial only if Dr G. selects correctly at the first and second trials; we saw (page 32) that the probability of his doing so is $\frac{1}{4} \times \frac{1}{4} = \frac{1}{16}$.

It is good practice, whenever one is able to do so, to establish a result by two different arguments. This provides at least a partial check that one has not made either a silly logical or arithmetical error.

Using similar arguments to those above we can easily verify that the probability of stopping at the end of the third trial is $\frac{3}{64}$, corresponding to success at the first and second trials and failure at the third. Extending the addition rule to three mutually exclusive events we get

$$Pr\text{(stopping at or before 3rd trial)} = \frac{3}{4} + \frac{3}{16} + \frac{3}{64}$$
$$= \frac{63}{64}.$$

Thus, the opposite event, continuing to a fourth trial, has probability

$$1 - \frac{63}{64} = \frac{1}{64}.$$

We may establish by similar lines of reasoning that

$$Pr\text{(success at all of four trials)} = \frac{1}{256}.$$

Now let's look at the relevance of these results to our experiment. The rules mean that we shall *accept* the hypothesis 'Gesser really is a guesser' if any one of a set of outcomes with a total associated probability (subject to the hypothesis being true) of $\frac{255}{256}$ is observed. We shall *reject* the hypothesis only if an event occurs which has a probability of $\frac{1}{256}$ when the hypothesis is true. Thus rejection implies either that our hypothesis is incorrect, or that in our particular experiment we have an outcome that we would only expect to occur in about one in every 256 experiments if the doctor really were guessing.

Formalizing our result

What we've just been doing is called a hypothesis test. There's plenty of theory and jargon associated with these tests, much of it over-worked, especially in scientific papers. One widely used concept is that of *significance*.

To explain it we need some more basic ideas. Given a hypothesis and a set of experimental results we may divide the collection of all possible outcomes of an experiment – often called the *sample space* or *record space* – into two mutually exclusive subspaces and calculate the probabilities assignable to each if the hypothesis we are testing is true.

For our experiment a possible breakdown of the sample space is given in Table 1. We accept the hypothesis of guessing if any of

Table 1 *Diagnosis experiment: events and their probabilities if guessing*

Event number	Description	Probability
1	Stop after first trial	$\frac{3}{4}$
2	Stop after second trial	$\frac{3}{16}$
3	Stop after third trial	$\frac{3}{64}$
4	Stop with failure at fourth trial	$\frac{3}{256}$
5	Stop with success at fourth trial	$\frac{1}{256}$

the events numbered 1 to 4 in this table occur, but reject it if event 5 occurs. The rule in the table divides the sample space into a *region of acceptance* (events 1 to 4) and a *region of rejection* (event 5). The region of acceptance contains the relatively more likely outcomes. The region of rejection contains the relatively less likely outcomes under the hypothesis upon which we based our probabilities; further, events (in our case only 1) in the region of rejection will have a higher probability under the alternative hypothesis that the doctor has some diagnostic ability. This is intuitively clear because he is more likely to be right if he has this ability.

Since the events 1 to 4 are mutually exclusive we may use the addition rule and add the probabilities of each of these events to obtain the total probability of a result in the region of acceptance. When we have more than one point in a region of rejection we may also calculate the total probability of a result in that region by the addition rule. Remember that the probabilities in Table 1 are *conditional* upon the hypothesis of guessing.

The region of rejection is often called the *critical region*. The total probability associated with the mutually exclusive events of which it is composed is the *size* of the critical region. If the outcome of an experiment is an event in the critical region we say the result is *significant*, that is, it provides worthwhile evidence that the hypothesis should be rejected. The size of the critical region specifies the *level of significance*. In our experiment this is $\frac{1}{256}$.

Why did we choose this particular region? This is a fair question for we might well choose a critical region composed of events 4 and 5 in Table 1. The size of this region is

$$\frac{3}{256} + \frac{1}{256} = \frac{1}{64}$$

and therefore $\frac{1}{64}$ is our new significance level. Note incidentally that with this choice of critical region there is no point in carrying out the fourth trial because, whatever its result, we reject the hypothesis that Dr G. just guessed. Thus the choice of significance level influences the size of the experiment we need.

Increasing the size of the critical region also increases the probability that we shall wrongly reject a hypothesis when it is in fact true – in this example from $\frac{1}{256}$ to $\frac{1}{64}$.

Except in certain trivial cases we can *never* prove or disprove a hypothesis in absolute terms. An example of a trivial case is provided by an experiment consisting of five tosses of a coin to test the hypothesis that it is double-headed. If one or more 'tails' are observed the hypothesis is immediately disproved.

In non-trivial situations a hypothesis test giving a result in the critical region leads to a statement of the form: 'Either the hypothesis is wrong, or in our experiment we have observed a result belonging to a set of outcomes that are unlikely under our hypothesis and more likely if some other hypothesis is true; the degree of unlikeliness is indicated by the significance level.'

Choice of a significance level requires thought. We do not want the associated probability to be too high. For example, a significance level of $\frac{1}{2}$ implies that we reject a hypothesis on results that would occur in about 50 per cent of our experiments even if the hypothesis was true. No rational person would want to do this.

In more complicated experiments convention plays a large part in choice of significance levels. In our experiment we had relatively few choices because of the somewhat restricted sample space. Indeed, if we felt $\frac{1}{256}$ were too small we had to jump to $\frac{1}{64}$ by incorporating one additional point in the critical region.

In more elaborate experiments one often has what amounts to a continuous choice of sizes for the critical region. Conventionally we do not usually consider critical regions of size greater than $\frac{1}{20}$, or as a decimal, 0.05. A commonly used expression is to refer to 'significance at probability level 0.05' when a result falls in a critical region of this size. The level is variously referred to as 'the 0.05 significance level', 'the 1 in 20 significance level' or 'the 5 per cent significance level'; all these mean the same.

Other widely used levels are 0.01 (= 1 in 100 = 1 per cent) and 0.001 (= 1 in 1000 = 0.1 per cent). The reasons for these conventions will become clearer in the two following chapters.

It is common practice to refer to results as *significant*, *highly significant* or *very highly significant* respectively when significant at the 5, 1 or 0.1 per cent level. Less desirably, these levels are sometimes indicated, especially in association with numerical results, by *, ** and *** – a notation that confuses rather than clarifies. Note that *higher* levels of significance, or more asterisks, correspond to *smaller* critical regions or *lower* probability values.

Our discussion so far has centred upon testing a single hypothesis. We have spoken only vaguely about accepting *some* alternative hypothesis if we reject it. Much of the literature on hypothesis-testing is framed in terms of a choice between two competing hypotheses. Not only does this have theoretical advantages but it also reflects reality. However, it may be difficult to put alternative hypotheses into probabilistic terms. How, for example, do we put into probabilistic terms the hypothesis that Dr G. has *some* (but not perfect) diagnostic ability?

Even if we reject the hypothesis that the good doctor is guessing we have still reached no conclusion about how successful he is at diagnosis, a practical question to which we would like an answer. It is a serious weakness of the hypothesis-testing approach that it does not answer this kind of question. There is another great weakness of the method, that we may accept a hypothesis when it is far from true.

Let's use our example to illustrate these points. Suppose we reject the hypothesis of guessing only if Dr G. succeeds at all four trials, i.e. using a $\frac{1}{256}$ significance level when testing the hypothesis $p = \frac{1}{4}$ (p is a useful shorthand for the probability of correct diagnosis). If the doctor has some, but imperfect, diagnostic ability the probability of his selecting the correct patient exceeds $\frac{1}{4}$, but is less than 1. Let us suppose that his ability is such that he selects the correct patient from a group of four in two out of every three trials in the long run. This implies that the probability of correct selection is $\frac{2}{3}$. In these circumstances the probability that he would select the correct patient in all four trials would be

$$\tfrac{2}{3} \times \tfrac{2}{3} \times \tfrac{2}{3} \times \tfrac{2}{3} = \tfrac{16}{81},$$

a fraction that is nearly equal to $\frac{1}{5}$. This implies that in about one experiment in five our result would fall in the critical region for which we would reject the hypothesis $p = \frac{1}{4}$ but in the remaining four experiments out of five (or more strictly 65 out of 81) we would accept the hypothesis $p = \frac{1}{4}$ when the *true* value of p was in fact $\frac{2}{3}$.

Increasing the size of the critical region gives some pay-off in that it reduces the chances of wrongly accepting a hypothesis. In our example if we include events 4 and 5 of Table 1 in the critical region we make its size $\frac{1}{64}$; then we can easily verify (try checking this yourself) that when $p = \frac{2}{3}$ the probability of a mistake occurring at or before the third trial is $\frac{19}{27}$. This implies that in these circumstances we would wrongly accept the hypothesis $p = \frac{1}{4}$ when in fact $p = \frac{2}{3}$ in about 19 experiments out of 27 – this is roughly two experiments out of three.

This unsatisfactory result underlines the fact that acceptance of a hypothesis by no means proves its correctness; it is wise to look upon acceptance merely as *non-rejection* because there is not sufficient evidence to conclude with conviction that the hypothesis is wrong.

Given time, one could study the behaviour of our test for a whole range of alternative values of p. More satisfactory is to specify a range of possible values of p with a reasonable degree of confidence that the unknown true value lay within that range. This range would depend upon our observed results. We shall return to statements of this type after we have developed some additional concepts. The first example of one is in Chapter 4.

A pause for questions

It's a little frustrating to develop an experiment and discuss its analysis only to find that there are a lot of shortcomings in the results. In hypothesis-testing we find in general that different experiments lead to different critical regions and that some experiments are better at discriminating between rival hypotheses.

The results of the experiment and analysis we have just been considering could most kindly be described as 'meaningful, but not useful'.

It's time for a critical examination of what we did. Why did we choose groups of four patients with only one diseased in each? Could we not have varied the numbers of diseased patients from group to group without telling the doctor the precise number he should find in each? Why did we restrict ourselves to four trials, and allow the doctor only one mistake?

These are valid questions but the statistician is likely to be evasive in answering them. People brought up on the clear-cut logical proofs and arguments of conventional mathematics sometimes find this irritating; but it is in the very nature of statistics that there must be compromises and provisos.

First let us make it clear that there is no magic about a group size of four, or about having only one diseased patient in each group. It would be as valid to have six patients in a group and to tell Dr G. that he may find between none and six diseased persons per group. This complicates the calculation of the probabilities associated with various events in the sample space and leads to different critical regions of appropriate sizes. With six patients per trial the total number of patients for four trials is also increased. If, on the other hand, we reduce the number of patients per group from four to three, less patients are required for four trials. Even if we adhere to one diseased patient per group the probabilities associated with the various outcomes will differ from those given in Table 1.

As an exercise you might like to show that if we reduce our group size to three, with one diseased patient per group, and reject the hypothesis of guessing only if the doctor selects the correct patient at all four trials, then the size of the critical region becomes $\frac{1}{81}$. You can use the multiplication rule to show this.

We may also relax the restriction to only four trials and the rule that the doctor is stopped after just one mistake. Will this make our experiment more realistic?

We might, for example, decide that the experiment will consist of eight trials, still with groups of four containing one diseased patient at each trial; however, instead of stopping as soon as a mistake is made we might allow Dr G. to complete all eight trials and base our decision to accept or reject on the total numbers of correct diagnoses and errors in all eight trials.

Perhaps this experiment might distinguish alternative hypotheses better. We saw that our earlier experiment would have led to acceptance of the hypothesis $p = \frac{1}{4}$ when in fact $p = \frac{2}{3}$ in about four performances out of five. The ability of a test to discriminate between hypotheses is called the *power* of the test. For our present purposes it suffices to say that we want a test to be powerful, in that there is only a small probability that we shall accept a hypothesis when it is not true.

Statisticians often compare several tests and different experimental set-ups which require roughly the same amount of resources and effort to select the best discriminator between alternative hypotheses of interest. In practice one set-up or one test procedure is not always best for a whole range of alternative hypotheses and some compromise is then called for.

The absence of clear-cut answers about the size and nature of the 'best' experiment is one reason why the statistician is part scientist, part craftsman. He employs his scientific skills when he uses probability concepts to work out the size of a critical region or the power of a test. He acts as a craftsman when he decides how to make the best use of limited available material and how to use vague information about alternative hypotheses and other aspects of his 'model' of a situation involving random variation.

He must also take into account the limitations which available resources place on the range of experiments possible.

We can clarify some of these points by looking further at our example. First is the problem of pinning down what we mean by 'some diagnostic ability'. We must always explain clearly what is meant by a phrase such as the 'probability of correct diagnosis is $\frac{3}{4}$'. The doctor may, for example, diagnose no one as diseased who is not diseased, but in the long run may only detect by his test three-quarters of all those who are diseased. A different situation is when only three-quarters of the patients the doctor designates as diseased are in fact diseased. The first situation could not in fact arise with the type of experiment we have been discussing because we require him to nominate one diseased person from each group. It is important to distinguish logically between the first kind of probability – that of detecting the disease when it is present – and the second type – the probability that a positive diagnosis is correct.

If the disease were serious we should prefer the doctor to spot all diseased patients even if he did worry a few healthy people with an incorrect diagnosis. That, however, is a problem of medical ethics rather than statistics.

To distinguish experimentally between the two probability situations we have just been discussing we could design an experiment in which the doctor was not told the precise number of diseased patients in each group.

However we do our testing, the hypothesis that the doctor is guessing is only a preliminary step. The practical question is 'If this method is any good at all, how does it compare with other systems of diagnosis?' This may involve not only questions about the number of correct results, but a comparison of the performance of this method with others. If Dr G.'s 'tongue' test is cheaper, quicker and nearly as good it might replace an established method. If it is less effective it might still be useful for a preliminary screening. By far the majority of laboratory and field experiments are *comparative* by nature and we shall meet examples of these. One type is examined in detail in Chapter 5.

An alternative experiment

Finally in this chapter let us look at an alternative experiment to try and establish whether or not the doctor is guessing. We shall compare its performance with the earlier one.

Again the doctor is presented with four patients, one of whom is diseased, at each trial. We still ask Dr G. to pick the diseased patient at each but do not stop him if he makes a mistake until he has completed eight trials, a set-up proposed on page 42.

We find that it is reasonable to reject the hypothesis that he is guessing even when he makes as many as 3 mistakes in 8 trials, no matter whether these be at the first three trials, at the first, sixth and eighth or in any other order. This experiment seems fairer to the doctor than one in which a mistake immediately disqualifies him.

In this fixed-size experiment of 8 trials the sample space can conveniently be divided into 9 mutually exclusive events corresponding to 0, 1, 2, 3, 4, 5, 6, 7 or 8 successes, each with its corresponding number of failures: for r successes (where r is any number between 0 and 8) there are $(8 - r)$ failures.

We can use the multiplication and addition laws to calculate the probabilities of each number of successes. The resulting probabilities are called *binomial probabilities*. 'Binomial' refers to the two possible outcomes of a trial – success or failure. For example, the probability of 8 successes is simply the product of the 8 probabilities of the independent events *success* at each trial, i.e.

$$Pr(8 \text{ successes}) = \tfrac{1}{4} \times \tfrac{1}{4} \times \tfrac{1}{4} \times \tfrac{1}{4} \times \tfrac{1}{4} \times \tfrac{1}{4} \times \tfrac{1}{4} \times \tfrac{1}{4}$$
$$= \tfrac{1}{65536},$$

when we accept the hypothesis $p = \tfrac{1}{4}$ corresponding to guessing.

The probability of 7 successes and 1 failure also involves the multiplication rule, and the addition rule as well. By the multiplication rule we first work out the probability of 7 successes and 1 failure in a given order. For instance, suppose that failure occurs (with probability $\tfrac{3}{4}$) at the sixth trial, then

Pr(failure at 6th trial, success at all other trials)
$$= \tfrac{1}{4} \times \tfrac{1}{4} \times \tfrac{1}{4} \times \tfrac{1}{4} \times \tfrac{1}{4} \times \tfrac{3}{4} \times \tfrac{1}{4} \times \tfrac{1}{4}$$
$$= \tfrac{3}{65536}.$$

If we allowed the one failure to take place at another trial, say the second, the probability of 7 successes and 1 failure would again be $\frac{3}{65536}$. Now there are 8 mutually exclusive ways of getting 1 failure and 7 successes, corresponding to the 8 different trials at which failure may occur. Thus we may add the 8 terms each of which has the value $\frac{3}{65536}$ to get the probability of 7 successes and 1 failure over-all. The required probability is thus $\frac{24}{65536}$.

The probabilities associated with other points in the sample space can be worked out in much the same way. The only difficulty is in working out the number of mutually exclusive events corresponding to different numbers, k, of failures. This is called the number of *combinations* of k items (failures) out of 8. We write this as 8_{c_k} where

$$8_{c_k} = \frac{8 \times 7 \times 6 \times \ldots \times (8 - (k - 1))}{1 \times 2 \times 3 \times \ldots \times k}$$

a standard result in algebra.

Thus, if $k = 3$,

$$8_{c_3} = \frac{8 \times 7 \times 6}{1 \times 2 \times 3} = 56.$$

The numerical values for probabilities of the various events, which look rather ugly, are given in Table 2.

Table 2 *Eight trials: events and their probabilities if guessing*

Event number	Description	Probability
1	0 successes, 8 failures	$\frac{6561}{65536}$
2	1 success, 7 failures	$\frac{17496}{65536}$
3	2 successes, 6 failures	$\frac{20412}{65536}$
4	3 successes, 5 failures	$\frac{13608}{65536}$
5	4 successes, 4 failures	$\frac{5670}{65536}$
6	5 successes, 3 failures	$\frac{1512}{65536}$
7	6 successes, 2 failures	$\frac{252}{65536}$
8	7 successes, 1 failure	$\frac{24}{65536}$
9	8 successes, 0 failures	$\frac{1}{65536}$

Clearly – as we would expect if the doctor were guessing – large numbers of successes have low probabilities, but these probabilities are increased if the true value of p is greater than $\frac{1}{4}$ because diagnostic ability is reflected in greater chances of success. So we form our critical region from those events in Table 1 that have lowest probability.

How large can the critical region be, subject to the condition that its size does not exceed $\frac{1}{20}$ – corresponding to the 5 per cent significance level? Clearly the region must include event number 9 because this has the lowest probability of all if $p = \frac{1}{4}$. A critical region including both events 8 and 9 has size

$$\frac{1}{65536} + \frac{24}{65536} = \frac{25}{65536}.$$

Similarly, a critical region composed of events 7, 8 and 9 has size $\frac{277}{65536}$. One composed of events 6, 7, 8 and 9 has size $\frac{1789}{65536}$. This is in fact the largest critical region we can form of size less than $\frac{1}{20}$, for if we add event number 5 to our region its size becomes $\frac{7459}{65536}$, which exceeds $\frac{1}{20}$. Our conclusion therefore is: with a significance level not greater than 5 per cent we may reject the hypothesis $p = \frac{1}{4}$ if 5 or more successes are observed.

In an earlier section we studied the performance of our hypothesis test when the true value of p was $\frac{2}{3}$. Calculations which you might like to have a go at yourself, but which we shall not give in detail, show that when p is indeed $\frac{2}{3}$ we would accept the hypothesis $p = \frac{1}{4}$ only in about 1 experiment out of 4.

We showed in the earlier experiment that a critical region of size $\frac{1}{64}$ (the largest available region of size less than $\frac{1}{20}$) in a similar situation leads to incorrect acceptance of the hypothesis $p = \frac{1}{4}$ in about 2 experiments out of 3. We have illustrated an important general principle: large experiments are usually more powerful than small ones, i.e. they discriminate better between hypotheses. We noted in our earlier example that using a smaller critical region (of size $\frac{1}{256}$) decreased our power of discrimination between the two hypotheses $p = \frac{1}{4}$ and $p = \frac{2}{3}$. This is another rather general result.

The reader who feels after all this that hypothesis tests are rather unsatisfactory has the right idea. For many practical problems they are not what is wanted. This does not mean we must dismiss them as completely useless. They do have a salutary effect by discouraging us from jumping to conclusions on too little evidence; but remember that a non-significant result, i.e. one for which we accept the hypothesis under test, may tell us only that the experiment is too small to be worthwhile.

An experienced statistician can give advice on the most suitable experiment for the circumstances. He bases his advice upon calculations that allow for various contingencies and upon his experience in similar situations. His aim is always to make the best use of available resources by minimizing the amount of experimental work needed, while remaining confident that he will be able to answer the questions the experimenter is asking.

Sequential experiments

Our original experiment, in which we stopped as soon as the doctor made a mistake, involved a decision about whether to stop or continue after the first, second and third trials. It provides a simple example of a *sequential* experiment. Sequential experiments usually have rather more complicated *stopping rules* – the rules that let us decide after each trial whether or not to stop, and if we do stop which of two specific hypotheses to accept. Although there is often an overall limit on the maximum number of trials, this is usually larger than the four of our example.

The advantage of sequential experiments is that they allow us to stop as soon as things look pretty clear cut; this is especially advantageous when observations by their nature come in one at a time.

In medical experiments they also have an ethical advantage. Once it is established beyond reasonable doubt that one course of action, say, a certain treatment, is more beneficial than another, then that more favourable action may be taken with all future patients.

Self diagnosis

Despite our disparaging attitude towards hypothesis-testing it does introduce some interesting statistical ideas. To make sure you've grasped the main ones try your hand at this little problem.

Instead of having groups of four at each trial the doctor is given groups of three, each with one diseased patient. He is allowed to examine a maximum of five groups. The hypothesis that he is guessing is rejected only if he makes no mistakes and picks the right patient in all five groups. What is the size of the critical region for this test? How is the size of the critical region altered if we reject the hypothesis of guessing if he makes none or exactly one mistake in the five trials?

If we reject the hypothesis of guessing only if he picks the right patient in all groups, what is the probability that we will accept the hypothesis of guessing when in fact there is a true probability of $\frac{3}{4}$ that he will select the right patient in any group?

If you are in difficulties turn to pages 223–4 for free diagnosis of your troubles.

3 Out for the Count

Categorical data

We pointed out in Chapter 1 that part of the strength of statistics lay in its ability to use the same mathematical models or techniques for data from widely different situations. A good illustration is provided by figures that consist of counts of items that can be categorized in two or more ways. We must be careful to see that our mathematical model gives a reasonable reflection of what is happening in the real-world problem; usually it is simplified, but it must mirror the important features.

The data in Table 3 are the results of an opinion poll of 244 electors. For illustrative purposes we have left out the 'don't knows', an omission which does not alter the basic principles of the analysis, although it does alter the practical interpretation.

Table 3 *Voting intentions of 244 electors*

	Conservative	*Labour*	*Liberal*	*Total*
U.K.-born	82	79	45	206
Immigrant	12	19	7	38
Total	94	98	52	244

The electors are categorized in the table by two characteristics, place of birth (U.K. or overseas) and political affiliation (Conservative, Labour or Liberal). It states, for example, that 79 out of 244 people in the sample were U.K.-born and intended to vote Labour, and that 7 were immigrants who intended to vote Liberal. Table 6 (page 61) and Table 8 (page 66) present two further sets of data amenable to analyses similar to the one we are about to apply to the data in Table 3. Note that mere similarity in appearance does not in itself always mean that data can or should be analysed in the same way.

We are going to use the data of Table 3 to answer the question: what evidence does the data provide that, in the electorate to which they refer, the percentage of people intending to vote for the three main parties is the same for U.K.-born electors as for migrants?

We can put the question another way by asking whether the data indicate that voting intentions are independent of place of birth as reflected by classification into U.K.-born and immigrant.

Another different form of the same question is to ask whether any difference in support for the various parties by people in the two birth-place categories, as calculated from the Table 3 data, could reasonably be attributed to sampling variation; that is, the 'luck' of whom we did or did not include in our sample.

In this situation we shall again resort to a hypothesis test – perhaps surprising after our lukewarm attitude towards these tests in the previous chapter. For this question a hypothesis test is appropriate.

The situation differs in several ways from the one in the diagnosis problem. There we set up the hypothesis that Dr Gesser was a guesser and a probability model was available before we even collected any data. Here we start from the data themselves and the hypothesis that birthplace does not affect voting intentions.

If the hypothesis is true figures for U.K.-born voters and immigrants are separate samples each of which gives information (subject to sampling variation) for estimating proportional support for the three political parties.

We can also suppose that if the hypothesis is true, then the totals in each column of Table 3 (U.K.-born + immigrant) provide better estimates of proportional support because they are based upon a larger sample.

We shall base our hypothesis test on a comparison of the proportional party vote for all electors in the sample with the proportions for each of the groups, U.K.-born and immigrant. Consistency between the groups would argue in favour of the hypothesis that voting intentions are independent of birthplace.

Before starting a formal analysis it is a good idea to examine the data critically to see whether they immediately suggest certain conclusions. If the results of a more formal analysis disagree violently with common sense we should consider whether we have selected the right mathematical model and check also for a mistake in our calculations.

Taking an informal look we see in Table 3 a slightly higher proportion of Labour voters among migrants than among U.K.-born electors. Our statistical analysis should tell us whether this could reasonably be ascribed to sampling variation or whether it suggests a *real* difference in the whole electorate for which our sample is 'representative'.

A down-to-earth mathematical model

We have already suggested that the same mathematical model can answer questions about the data in Tables 3, 6 and 8. Although we shall develop our model specifically using the numbers for the example on voting intentions, its general application will be evident.

Suppose we put 244 tickets in a hat, one corresponding to each of our 244 electors. Let us label 94 of these Conservative, 98 Labour and 52 Liberal, corresponding to the party support given by the column totals. If we now mix the tickets thoroughly and draw out 38 (corresponding to the total number of immigrants) we have a *random sample* of 38 from the 244 tickets. The idea of a random sample is very important in statistics and we shall return to it again and again. It has the essential property in our case that every ticket has exactly the same chance of being included in the sample, no matter what party label is attached to it.

Our sample of 38 turns out to be 12 Conservative, 19 Labour and 7 Liberal. Given the total number of tickets with each party label, should this result surprise us?

Well, what do you think? We would certainly have been surprised if our sample of 38 had contained no Conservatives, 2 Labour and 36 Liberals. It is possible to get such a random sample but the odds against it are very long indeed. A sample of 38 Conservatives and no representatives of the other parties is again possible, but the odds against it even longer. It is extremely unlikely that if the tickets were properly mixed that we would not draw out at least one of the 150 tickets *not* marked Conservative in 38 draws.

Although it would be quite time-consuming even with a modern computer, we can work out the probability of getting each and every conceivable sample of 38 out of 244 tickets with the numbers of party labels set out above. The 206 tickets left in the hat would then provide a sample corresponding to the 206 U.K.-born electors.

It is intuitively clear, and can be verified either theoretically or experimentally, that random samples of 38 with proportions of labels similar to those in the total sample of 244 are more likely than are samples with proportions far removed.

We can easily work out the *expected numbers* of voters for each party in a sample of 38 when we know the proportions for all 244. The expected numbers are simply those that give the same proportions in the sample of 38 as among all 244 electors. If we have 94 tickets marked Conservative then we *expect* $\frac{94}{244}$ of our 38 to be Conservative, i.e. the expected number of Conservatives is

$$\frac{94}{244} \times 38 = 14.6.$$

Similarly, the expected number of Labour voters is

$$\frac{98}{244} \times 38 = 15.3,$$

and of Liberals

$$\frac{52}{244} \times 38 = 8.1.$$

Clearly we can only observe an *exact* number of voters for each party so we shall never observe precisely the expected numbers, but outcomes near the expected numbers, such as 15:15:8 or 16:17:5, should certainly not surprise us.

One way to test our hypothesis of no association between birth-place and voting intentions is to set up a critical region as we did in Chapter 2 and to reject the no-association hypothesis if our experiment gives a result in this critical region. The appropriate critical region consists of results that are highly unlikely if there is no association, but which are more likely if some association exists. If our observed sample did not fall in this critical region we would continue to accept the no-association hypothesis. The difficulty with this approach is working out the probabilities associated with a vast array of results that *might* have occurred but in fact did not. We need these to determine the critical region.

Fortunately there exists a short-cut approximate test that does essentially the same thing. The short-cut test takes quite a deal of mathematics to justify theoretically, but we can show it is intuitively reasonable. To use it we need the expected numbers in each category in the body of Table 3. We have worked these out already for the immigrants. For the U.K.-born (corresponding to tickets left in the hat) these are

Conservative $\frac{94}{244} \times 206 = 79.4$,

Labour $\quad\quad \frac{98}{244} \times 206 = 82.7$,

Liberal $\quad\quad \frac{52}{244} \times 206 = \underline{43.9}$

Total $\quad\quad\quad\quad\quad\quad\quad\quad\quad 206.0$

To follow the well-established principle that it is wise to check a calculation by performing it in a different way, we note that we can deduce the expected numbers for all remaining categories after having calculated only two of them because the *expected* numbers in any row or column add up to the *observed* total for that row or column. For example, if we had the expected numbers 14.6 and 15.3 for Conservative and Labour among immigrants, then, since there are 38 immigrants, the expected number of Liberals among immigrants is

$$38 - 15.3 - 14.6 = 8.1,$$

agreeing with the value obtained directly. Also, since there are 94 Conservatives in all and since the expected number amongst immigrants is 14.6, that among U.K.-born electors must be

$$94 - 14.6 = 79.4.$$

Similarly, among U.K.-born electors the number of supporters for Labour are

$$98 - 15.3 = 82.7,$$

and for Liberal are

$$52 - 8.1 = 43.9.$$

The short-cut test of the no-association hypothesis hinges on the fact that small discrepancies between observed and expected numbers in the various categories support the no-association hypothesis while large discrepancies argue against it. One simple idea is to add up the discrepancies

$$observed\ number\ -\ expected\ number$$

for each entry in the table. It is convenient to call each position in the table a *cell*. So that we may readily compare observed and expected numbers in each cell we give in Table 4 the expected numbers corresponding to the observed numbers in Table 3. The discrepancies are easily calculated and we find their sum to be

$$(82 - 79.4) + (79 - 82.7) + (45 - 43.9) + (12 - 14.6) + (19 - 15.3) + (7 - 8.1)$$
$$= 2.6 - 3.7 + 1.1 - 2.6 + 3.7 - 1.1$$
$$= 0$$

This is not a very useful result!

Table 4 *Voting intentions: expected numbers*

	Conservative	Labour	Liberal
U.K.-born	79.4	82.7	43.9
Immigrant	14.6	15.3	8.1

A moment's reflection shows that our real interest is in the magnitudes of the discrepancies irrespective of their directions (i.e. whether they are positive or negative), since a negative value of

$$observed\ number\ -\ expected\ number$$

indicates departure from independence just as much as the same positive value. It is thus more logical to use for each cell some function of the discrepancy that is always positive. An obvious candidate is the *magnitude*, i.e. the discrepancy taken as a positive value irrespective of its direction. For purely technical reasons – chiefly because the mathematical theory is easier – statisticians prefer to use the squares of the discrepancies, i.e.

$$(observed\ number\ -\ expected\ number)^2.$$

The larger the discrepancies the larger the sum of these squares over all cells. But a discrepancy of a certain magnitude looks less alarming with a high expected number than if the expected number is low. This suggests that it is more appropriate to consider the square of each discrepancy as a fraction of the expected number for the corresponding cell and to calculate

$$\frac{(observed \; number \; - \; expected \; number)^2}{expected \; number}$$

and to add the results for all cells to get an overall measure of discrepancy. A high value is termed significant and indicates a departure from the no-association hypothesis.

How high is high? Fortunately tables now come to our aid. The tables used are the *chi-squared tables*, often written as the χ^2 tables. The symbol χ is a Greek letter spelt 'chi' and pronounced 'ki'. The quantity we calculate is called the *chi-squared statistic*. The word *statistic* – singular of statistics – denotes a number calculated from a set of data.

Degrees of freedom

In obtaining our chi-squared statistic we worked out the expected numbers in each cell. To calculate these we used the known marginal totals. The value of the chi-squared statistic depends only upon the observations through the differences between these and the expected numbers.

Now we noted on page 54 that the sum of these differences was 0. The differences have another striking feature, namely that the sum of the differences for the cells in any row or column in the tables is also 0. So if we know two of the differences, provided they are not in the same column, we can work out all the discrepancies between observed and expected values. For example, for U.K.-born Conservative voters the difference is

$$82 - 79.4 = 2.6,$$

while for U.K.-born Labour voters it is

$$79 - 82.7 = -3.7.$$

These are both in the first row and since the sum of the discrepancies for the row is 0, the remaining discrepancy must be 1.1. It is easily verified that this is correct. Similarly, since the column differences total 0 the '*observed — expected*' values in the second row are clearly —2.6, 3.7 and —1.1. We have also seen on page 53 that we only needed two of the expected numbers (provided they were in different columns) to work out the remainder because the expected numbers in any row or column add to the observed totals (the *marginal totals*) for that row or column. Thus given our observations – which determine the marginal totals – we only need two *pieces of information* about discrepancies to work out the chi-squared statistic. This is the minimum requirement – we could not work out chi-squared from only *one* expected number. This minimum number of pieces of information required to work out a statistic from a set of observations is called the *degrees of freedom* of the statistic.

We can look at degrees of freedom in a slightly different way. Our initial observations consist of six independent pieces of information – the observed numbers in each cell of Table 3. To form our chi-squared statistic we have to use some of the information. Specifically, we use the grand total and the marginal totals for rows and columns to calculate the expected numbers. This involves a minimum of four bits of information, which could be the grand total of all observations, one row total and two column totals. Given these we can calculate the other row total and the remaining column total by subtracting marginal totals from the grand total. We have now used four of our six original pieces of information to fix the marginal totals needed to calculate the expected numbers. This leaves two of the original six free and we again refer to two as the number of degrees of freedom.

In passing note that the choice of the grand total, one row total and two column totals as the four pieces of information that determine all marginal totals is not unique. The important point is that the number four is a minimum. We could deduce all totals given, for example, *either* both row totals and any two column totals *or* all three column totals and one row total. In each case we need at least four marginal totals from which to deduce the others.

In summary then, our initial six observations represent six degrees of freedom. Four of these are used when we fix the marginal totals needed to calculate the expected numbers, leaving two degrees of freedom for the statistic.

Using our tables

Table 5 is an abbreviated chi-squared table. More elaborate ones will be found in many statistical text books and in collections of tables such as Lindley and Miller (1970) or Fisher and Yates (1963). Tables vary slightly in layout but the usual pattern is for each row to refer to a specific number of degrees of freedom and each column to a probability level or the corresponding percentage. We are only interested in probability levels 0.05, 0.01 and 0.001 which correspond to 5, 1 and 0.1 per cent significance levels. The degrees of freedom vary from problem to problem. Our example had two.

Table 5 *Minimum chi-squared values for significance at indicated levels*

Degrees of freedom	Significance level %		
	5	1	0.1
1	3.84	6.63	10.83
2	5.99	9.21	13.81
3	7.81	11.34	16.27
4	9.49	13.28	18.47
6	12.59	16.81	22.46
10	18.31	23.21	29.59

The entries in the table are the minimum or critical values indicating significance at each level for the degrees of freedom indicated. It requires lengthy computations to obtain these values for any chosen probability level or degrees of freedom. This is one of the reasons, indeed the main reason, why we stick to the conventional 5, 1 and 0.1 per cent significance levels and so avoid very bulky tables.

In the computer era this excuse for sticking to convention has worn rather thin for it is now only a matter of seconds for a properly programmed computer to tell us at exactly what level any particular value of a chi-squared statistic would be 'significant'. Nevertheless tables are still widely used and are very convenient if one does not have first-class computing facilities at one's finger tips.

Let us see how we use the tables for our example. Denoting by χ^2 our statistic calculated by forming

$$\frac{(observed\ number\ -\ expected\ number)^2}{expected\ number}$$

for each cell and totalling over all cells we get

$$\chi^2 = \frac{(2.6)^2}{79.4} + \frac{(3.7)^2}{82.7} + \frac{(1.1)^2}{43.9} + \frac{(2.6)^2}{14.6} + \frac{(3.7)^2}{15.3} + \frac{(1.1)^2}{8.1}$$
$$= 1.79.$$

Table 5 tells us that with two degrees of freedom χ^2 must exceed 5.99 for significance at the 5 per cent level. Our observed value of 1.79 is well below this, so we do *not* reject the no-association hypothesis. Note carefully that we have not proved an absence of association; we have only arrived at a considered judgement that we have insufficient evidence to establish an association.

The test and its meaning

We have performed a test and come up with a non-significant result implying that in the *population from which the sample was taken* there is no satisfactory evidence of association between voting intentions and birthplace. This information is not of much use unless we know what that population is. Before we can define the 'relevant' range of any inferences we must know how the data were collected. In our example we may ask whether the data represent a particular electoral division. In the United Kingdom the country is divided into electorates, each of which returns one member of parliament. We must ask whether data are typical of one electorate only or of the whole country.

If the data referred only to workers in one factory or in one town, or to members of one socio-economic class, it would be ridiculous to use them to infer anything useful about nation-wide voting intentions. Trite and obvious, you may well say, yet numerous examples can be found in newspapers, or in radio and television broadcasts where people make sweeping assertions based upon too little or irrelevant information. On the same day the same set of economic returns is used to forecast both boom and doom. This sort of bad use of statistics understandably makes the public cynical. It is not the data alone that have led to opposite conclusions in such cases, but their interpretation by people of differing opinions; interpretation often arrived at by associating this data with other sets of information. All too often we are not told what these different opinions or extra pieces of information are.

Suppose the data in Table 3 (page 49) had all been collected in an electoral division where the percentage votes for each party at the last election were almost identical with those for the nation as a whole, as were the proportion of the electors who were immigrants. Many people given this information would regard the electorate as typical and if assured that the sample of 244 was *representative* would be prepared to use the data in Table 3 to make inferences for the whole nation.

This word *representative* calls for comment. A representative sample must include, in approximately the same proportions as in the population from which it is taken, people of various ages, occupations, economic status, religions, sexes and any other classificatory factor that might influence voting intentions.

Here we meet the practical difficulty that a very large set of factors – some perhaps unknown to us – influences voting intentions. Rather than trying to balance the sample for all of these it is more realistic to take a *random* sample: basically this requires that every individual in the electorate has the same chance as every other individual of being included in the sample. Again, selection of a truly random sample is a non-trivial process and one often settles instead for a *pseudo-random sample*: one that behaves for all practical purposes as though it were a random sample.

Here's how we might get a pseudo-random sample of voters from an electorate. If our sample were to be 1 per cent of the electorate we might select every 100th name from the electoral register starting with a randomly selected name in the first 100. Although this would not be truly random there is no reason to suppose that choosing every 100th name would make the sample unrepresentative.

There are several variations on the theme of simple random sampling, some quite sophisticated, to ensure more representative samples. The reader interested in pursuing this topic should consult Yates (1960) or Cochran (1963).

One instance where random samples are often avoided is in opinion polls; they use samples instead of a rather specialized form called *quota samples*. Quota sampling has certain economic advantages for organizations conducting polls and if used skilfully gives acceptable results, although it is sometimes hard to check their accuracy.

Even if the data in Table 3 were an effectively random sample from an electorate that had previously shown voting and birthplace patterns similar to the nation as a whole, we must still be cautious in making inferences from the sample about future national voting patterns. However typical an electorate may once have been, it may no longer be so. New factories employing large numbers of craftsmen may have swamped the rural section of the electorate; a local member of parliament may have made himself thoroughly unpopular since the last election.

The above points give some idea why it is important to know the source of data and show why we must take account of relevant supplementary information (like changes in electorate composition) when making inferences.

A different problem using the same mathematical model

Let's compare two drugs for treating a disease. Suppose we start with 27 sufferers. We might treat 15 with drug *A*, the remaining 12 with drug *B*. The results of such an experiment are given in Table 6.

If we suffered from the disease the figures in Table 6 might encourage us to hope that our doctor would prescribe drug *A*.

Table 6 *Responses of patients to two drugs*

	Recovery	Death	Total
Drug A	10	5	15
Drug B	6	6	12
Total	16	11	27

Before we pin too much hope on this it might be wise to seek answers to three questions:

1. Were all factors in patient care, other than the drug prescribed, the same for all patients; e.g. were they all in the same hospital, did they receive the same nursing care, were they all equally ill when treatment started? If not, then

2. was any attempt made to balance these factors so that they did not favour the group receiving one particular drug? Apart from this,

3. were patients allocated to each drug so as to eliminate any subjective element such as doctors giving a particular drug to patients more likely to recover irrespective of treatment?

Even when the answers to these questions indicate that the drugs are being compared on a fair basis we still have to contend with sampling variation; by this we mean that if we repeated the experiment on further groups we would be unlikely to get exactly the same responses.

To test a hypothesis of no association between the drug administered and recovery rate we can use a similar model to the one we used for voters. From the marginal totals in Table 6 we work out that if there is no association the expected number to recover when drug A is used is

$$\tfrac{16}{27} \times 15 = 8.9.$$

Given this expected number and the marginal totals we can work out all the remaining expected numbers by subtraction; e.g. the number expected to die amongst those treated with drug A is

$$15 - 8.9 = 6.1,$$

and as a check we may calculate this directly as

$$\tfrac{11}{27} \times 15 = 6.1.$$

Similarly, the expected number to recover under drug B is

$16 - 8.9 = 7.1$,

and to die

$11 - 6.1 = 4.9$.

As a final check we note that

$7.1 + 4.9 = 12$,

which is the correct marginal total for drug B.

We had to work out only one of the expected numbers *ab initio* from the marginal totals; thus there is one degree of freedom. For each cell the difference between observed and expected number is of magnitude 1.1. It is characteristic of tables with only two rows and two columns – often called 2×2 (pronounced 'two by two') tables – that the differences have the same magnitude for all cells since differences for each row and column must total zero. We now calculate the statistic

$$\chi^2 = \frac{(1.1)^2}{8.9} + \frac{(1.1)^2}{6.1} + \frac{(1.1)^2}{7.1} + \frac{(1.1)^2}{4.9}$$
$$= 0.75,$$

a value well below the 3.84 that Table 5 tells us is required for significance at the 5 per cent level with one degree of freedom. We therefore cannot reject the hypothesis of no association between drug administered and cure.

The more the merrier

How happy can we be about this result?

Looking again at Table 6 we see that of those receiving drug A $\frac{2}{3}$ recover while $\frac{1}{3}$ die. For drug B only $\frac{1}{2}$ recover, yet our test shows no firm base for accepting superiority of drug A. It seems intuitively reasonable to hope that if the same trend continued in a larger experiment, that is, the proportions recovering and dying remained unaltered, then we might establish the superiority of drug A.

Suppose we examined ten times as many patients and found the same proportions recovering and dying; we should then have the situation of Table 7.

Table 7 *More patients – same proportional responses*

	Recovery	Death	Total
Drug A	100	50	150
Drug B	60	60	120
Total	160	110	270

The expected numbers are now all ten times as great as before. Working to one decimal place they are 88.9, 61.1, 71.1 and 48.9 whence

$$\chi^2 = \frac{(11.1)^2}{88.9} + \frac{(11.1)^2}{61.1} + \frac{(11.1)^2}{71.1} + \frac{(11.1)^2}{48.9}$$
$$= 7.66.$$

This exceeds the value of 6.63 which Table 5 shows is required for significance at the 1 per cent level with 1 degree of freedom. So the larger experiment has indicated highly significant differences in the performances of the drugs although the percentage cure with each drug is the same as it had been in the first experiment.

We have again illustrated something encountered in the previous chapter: increasing the size of an experiment may enable us to reject a hypothesis that is not rejected by a smaller experiment, even though the pattern of results may be similar. The reason for this is the sampling variation which may well mask real effects in a small experiment. This is consistent with our intuitive ideas. If we toss a coin 4 times we are not surprised to get 3 heads and 1 tail; if we toss it 400 times we would be very surprised to get 300 heads and 100 tails. Yet in each case the result is heads 75 per cent of the time. Here is a timely warning of the danger of quoting percentages – a misleading concept without an indication of the sample size upon which they are based.

Although a hypothesis test tells us something we again see its limitations in a drug experiment. What we really want to know is whether our *sample* result of a ⅔ cure rate with drug *A* and a ½ cure rate with drug *B* really reflects the *population* difference. The population here is the total of diseased patients, present and future, who might be treated with the drugs used in this experiment and of which the observed patients can be regarded as a representative sample.

Ideally we should know whether our observed cure rate for drug A of $\frac{2}{3}$ indicates a population cure rate almost certainly between, say, 60 and 70 per cent, or whether the experimental evidence is only strong enough to define the rate as between, say, 50 and 90 per cent. This problem is basically one of estimation and can be dealt with by the appropriate theory. We shall leave our discussion of estimation problems to Chapter 4, where it is easier to build up a link between problems of estimation and hypothesis testing.

Chi-squared: some warnings

Care is needed with chi-squared tests. They are used as approximate tests in several different contexts where they take slightly different forms. One must be sure what question one is trying to answer. For example, suppose we had the data in Table 3 and also information on the proportional vote for the three parties for the whole of the United Kingdom at the last election. We might then like to know whether the data in the table suggest that the electorate to which they refer reflects the national voting pattern at the last election, or more particularly whether that pattern is reflected both by U.K.-born and by immigrant electors in that electorate. A chi-squared test could still be used, the questions are different, and we have additional information. The form of the test is therefore different.

An eminent statistician claimed that the way to get a significant result from any data was to have them analysed by enough statisticians – one was bound to find something significant. This would almost certainly be true if each tested a different hypothesis, made different assumptions or used other additional information. It is important to specify clearly what questions one is asking before attempting *any* statistical analysis and for the statistician to ensure that the question he is answering is the one the person seeking his advice is asking. The statistician also has a duty to 'come clean' about any assumptions he is making when he provides an answer. This is why all experimental work that involves statistics requires team work between data collector and data analyst.

We have pointed out that the chi-squared test is usually an approximation. Some conditions must be satisfied if it is to be a good approximation. Theoretically the approximation is very good if the *expected* numbers in *all* cells of a table are *large*. However, practical experience rather than complicated mathematical theory has also shown that the approximation works well if *none* of the expected numbers is *very small*. A convenient rule of thumb is: the chi-squared test for lack of association of attributes may be used if the lowest expected number for any cell is 5 or more. This is a useful guide but not the hard-and-fast rule some text books imply.

What if the expected number in any cell is less than 5? Sometimes a more tedious exact test must be used. In other cases it may be permissible to pool results for two or more cells to build up the expected numbers. The price one pays for this is a reduction in the number of degrees of freedom. Table 5 does not immediately show how we pay that price for reducing the degrees of freedom, but the fewer the number of discrepancies the fewer the degrees of freedom, and the value of the chi-squared statistic for significance at any given level is relatively higher *per degree of freedom* when these are small in number.

Why we have difficulties with small expected numbers becomes clear from a simple numerical example. Suppose that in a cell we have an observed number of 8. The discrepancy is 5 when the expected number is 3 or when it is 13. In the former case the contribution of that cell to the chi-squared statistic is

$$\frac{(8 - 3)^2}{3} = 8.33,$$

while in the latter case it is

$$\frac{(8 - 13)^2}{13} = 1.92.$$

For very small expected numbers the chi-squared test tends to give too many significant results, i.e. more than we would get with an exact test.

The chi-squared test also tends to give too many significant results for the 2 × 2 table. The approximation is improved for these tables if we apply *Yates's correction*: subtracting $\frac{1}{2}$ from the magnitude of each discrepancy between observed and expected number before we square it. This reduces the value of chi-squared. We need only use this correction when the expected numbers are not very large or the chi-squared statistic is close to the value required for significance.

A do-it-yourself chi-squared example

Table 8 gives the numbers of grade *A*, *B* and *C* bananas in samples of 100 from each of four shipments.

Table 8 *Banana grades: samples of 100 from each of four Shipments*

Shipment	Grade		
	A	B	C
1	22	65	13
2	29	62	9
3	14	70	16
4	15	71	14

Try calculating the appropriate chi-squared statistic to see whether the proportions in the grades differ significantly from one shipment to another. Have a good try, but if you are really stuck turn to Chapter 13, page 224.

The figures in this example have been 'cooked' to make the arithmetic easy. Real-life data aren't usually so kind to the statistician.

Now we have used the same mathematical model and the same statistical test for problems to do with voting, treating disease and grading fruit – not a bad cross-section of situations.

4 The Analysis of Measurements

Hard times ahead

This is a chapter where we look at many new ideas, all brought together because they can readily be discussed using data from one experiment.

In the last chapter the data all consisted of counts, numbers of individuals or items – voters, patients or bananas – with various characteristics. Here we look at measurements and see how we use them to answer several questions. In this and the next chapter new ideas come rather thick and fast, but they illustrate many points important to statisticians in their everyday work.

An experiment on visual response times

The data in Table 9 (page 68) are taken from a paper by Geffen, Bradshaw and Nettleton (1973) in the *Quarterly Journal of Experimental Psychology*.

They were exploring whether certain numbers presented in random order to individuals were perceived more rapidly in the right or left visual field – or whether there was no consistent difference, it being a matter of chance whether an individual responded more rapidly in one field or the other.

At five-second intervals a number – either 1, 2, 4 or 5 – was flashed on a display tube for 150 milliseconds (ms.). There was a tube in the right visual field (RVF) and one in the left (LVF). The subject was asked to say 'bonk' when either of two specified numbers appeared, but to say nothing if one of the other numbers were displayed. Each of the four numbers appeared exactly twenty times in pseudo-random order on each tube.

The paper gives a full and clear account of how the experiment was conducted. Precautions were taken to eliminate head movement that might result in information being transmitted to the incorrect field. All twelve participants were right-handed and had normal speech and vision; none had any close relatives who were left-handed. These requirements clearly remove potential sources of 'noise' or uncontrolled variability, but they restrict the validity of any inferences we make to a population that excludes people with left-handedness, or vision or speech defects.

The subjects were paid volunteers. This might be relevant if we believed that the lure of payment might attract people atypical of the population in which we are interested. Impecunious students, for example, might be more alert and respond more quickly than that mythical being 'the average man'; or the opposite might be true, their responses being slower. Yet another experiment would be needed to find this out.

Table 9   *Mean response times to digital information (milliseconds)*

Subject	Left visual field (LVF)	Right visual field (RVF)	Difference LVF − RVF
1	564	557	7
2	521	505	16
3	495	465	30
4	564	562	2
5	560	544	16
6	481	448	33
7	545	531	14
8	478	458	20
9	580	560	20
10	484	485	−1
11	539	520	19
12	467	445	22
		Total	198
		Mean	16.5

Table 9 gives for each individual in each visual field the average of the forty response times to the two relevant numbers and also the differences between these, in the column headed (LVF − RVF). Exceptionally long or missed responses, amounting to about 5 per cent of the total, were disregarded. We must be a little cautious about this omission because there is an obvious subjective element in deciding when a response time is 'excessively long'. An important practical problem in statistics is deciding when an observation should be rejected as non-typical or erroneous (more about that on page 88).

In this experiment the researchers wisely tested if the number of missed responses differed between fields. The test did not reject the hypothesis that the number of missed responses were independent of the field (but remember that non-rejection of a hypothesis is not proof of its truth). The test used by the authors for this depended on the chi-squared statistic, our friend from Chapter 3, here used in a rather different context.

Presenting the numbers in pseudo-random order was intended to avoid the possibility of any learning effect leading to recognition of a pattern. Strictly speaking, randomization is nearly always a precondition for a valid statistical analysis; nevertheless, many analyses are performed without this precondition. To decide whether this is justified in a particular case the statistician must combine the art and science of his calling.

Making a correct decision is very much a matter of practical experience. C. P. Cox (1968) put it well when he said 'the answer is not always an insistence on the strict letters of the law, rather that we should perhaps require an appreciation of the risks of deviating from them'.

The consequences of failing to randomize may vary from trivial to disastrous. Even if the need for randomization is not always obvious, its use is a wise precaution.

Another important concept illustrated by this experiment is that of *replication* or repetition on more than one unit; replication has already been met with the concept of repeated trials in Chapter 2. In the current example there are twelve replicates, each being represented by one individual or subject. Replication gives us a way of measuring natural variation between similarly treated units. This variation is the unwanted noise in our experiment, as distinct from what we are really interested in, the average difference between response times for each field.

Although noise has much to commend it as a term, statisticians more commonly refer to error variation, often, as we have already mentioned, inappropriate jargon in biological contexts where such variation is quite natural.

One interest to the experimenters was whether their results supported a theory of no consistent difference between *mean* or *average* (both mean the same) response times for left and right visual fields in the population for which our twelve subjects could be regarded as a sample. Population here is the somewhat nebulous concept of all people except those with peculiarities likely to affect their response times.

A quick and dirty test

It is good statistical practice to look intelligently at data and sometimes to try simple tests prior to a more formal analysis. Unhappily, we cannot do this when modern data processing methods do not let us see the figures before they are gobbled up by a computer. Automatic recording devices may put data directly onto punched cards or magnetic tape ready for the computer, where they are quickly processed by some (not always appropriate) program. Contrariwise, even if they are available, examining large bodies of data visually prior to some condensation or summarization is a formidable task. In Chapter 11 we shall see how the computer can be an ally in such matters.

Fortunately we have not got these problems with our visual field data, so let's take a look at them. From the second (LVF) and third (RVF) columns in Table 9 we easily see that the variation between individuals within one field (either left or right) is greater than that between the left and right fields for any individual. For example, in the LVF the highest individual response time is 580 ms. for subject 9, and the lowest is 467 ms. for subject 12, a difference of 113 ms. There is a generally comparable range of response times between individuals for the RVF.

Column four of the table gives the difference (LVF − RVF) for each individual. These range only from −1 (subject 10) to 33 (subject 6). Another important feature of the differences is that all but one are positive, implying that mean response times are generally longer in the LVF.

Figure 3 shows these differences graphically. The lengths of the vertical bars represents the magnitude of the differences (LVF − RVF) for each subject. Bars above the horizontal axis represent positive differences and the only one below, the negative difference.

As we have already suggested, the simplest question the experimenter might want to answer is 'do the data imply a consistent difference in mean response times for the two fields?' We can find an answer using the very simple *sign test* which is based, as its name implies, on the signs of the observed differences.

In formal terms the statistician may regard answering the question about differences in population mean response as a test of this hypothesis: the differences in individual response times for the fields have population mean zero. The rationale of the sign test is as follows. If the population mean difference (LVF − RVF) is zero, and it is just a matter of chance in which field the individual responded quicker, then we expect about half of a

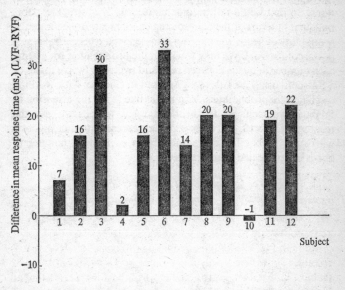

Figure 3 Differences in mean response times between the two visual fields

sample to give a positive value for (LVF — RVF) and about half to give a negative value. Examining the signs is analogous to looking at the numbers of heads and tails recorded when a true coin is tossed.

In the sign test positive differences are equivalent to heads and negative differences to tails. No difference between mean response times for the two fields is implied if we get nearly the same numbers of positive and negative differences. We have observed eleven plus signs and one minus sign – equivalent to eleven heads and one tail. What are we to conclude?

The analogy with the coin-tossing experiment could be made for many other experimental situations. For example we can give different diets to each lamb in a number of pairs of twins and wish to test if the population mean effects of the two diets – as measured by weight gain over a fixed period – can reasonably be assumed to be equal.

We now need a mathematical model to describe all such situations. So that we do not become unnecessarily abstract let's formulate it in terms of the coin-tossing experiment. If p denotes the probability of heads when a coin is tossed then p takes the value $\frac{1}{2}$ (if the coin is true). We may work out the probabilities for getting each number of heads between zero and twelve in twelve tosses of a true coin using the rules we discussed in Chapter 2 for working out binomial probabilities. You might like to try the extension to this case yourself. As the arithmetic gets tedious we give the results in Table 10, and their pictorial representation in Figure 4 where the probability corresponding to each given number of heads is represented by a vertical bar.

Table 10 *Heads and probabilities: twelve tosses of a coin*

Number of heads	Probability	Number of heads	Probability
0	$\frac{1}{4096}$	7	$\frac{792}{4096}$
1	$\frac{12}{4096}$	8	$\frac{495}{4096}$
2	$\frac{66}{4096}$	9	$\frac{220}{4096}$
3	$\frac{220}{4096}$	10	$\frac{66}{4096}$
4	$\frac{495}{4096}$	11	$\frac{12}{4096}$
5	$\frac{792}{4096}$	12	$\frac{1}{4096}$
6	$\frac{924}{4096}$		

In Table 10 the sum of the probabilities associated with the mutually exclusive and exhaustive set of events that are represented is 1, as it should be. Remember that we showed in Chapter 2 that we use the addition rule for mutually exclusive events and the multiplication rule for independent events. Note also in Table 10 the equality of the probabilities for 0 and 12, for 1 and 11, for 2 and 10 heads, etc. This is also evident in Figure 4.

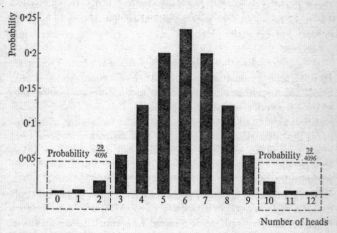

Figure 4 Probability of each number of heads in twelve tosses

What about the critical region for testing the hypothesis $p = \frac{1}{2}$? We determine it just as we did in Chapter 2 (page 46). First we look at the two extreme results, 0 heads or 12 heads. A critical region based on these outcomes has size

$$\frac{1}{4096} + \frac{1}{4096} = \frac{1}{2048}.$$

If we add to the critical region the outcomes 1 or 11 heads its size becomes

$$\frac{1}{2048} + \frac{12}{4096} + \frac{12}{4096} = \frac{26}{4096}.$$

If we also include 2 or 10 heads the size becomes $\frac{158}{4096}$ (you should verify this) which is still less than the conventional upper limit of $\frac{1}{20}$ for a critical region.

We can easily verify that if we add the probabilities corresponding to 3 or 9 heads the size is above $\frac{1}{20}$; this would create a region larger than the conventionally acceptable size. Therefore the critical region we use will consist of the sample points 0, 1, 2, 10, 11, 12 in the dotted rectangles of Figure 4.

From the values in Table 10 and our calculations to determine the appropriate critical region, we see that rejecting the hypothesis that the coin were true if we observed either 0 or 1 head or 11 or 12 heads in fact uses a critical region of size $\frac{26}{4096}$. This corresponds to a significance level well below the conventional 1 per cent.

We can translate these ideas to the visual response experiment. In it we have 11 positive and 1 negative difference, so we should reject the hypothesis of population mean difference being zero at the 1 per cent significance level.

All this is fine so far as it goes, but it tells us nothing about *how big* is the population mean response time difference. Indeed it has made very little use of the information our data contains; it has looked only at the signs of the differences, and ignored their magnitudes.

The sign test is an example of a general class of tests usually called *non-parametric* or *distribution free*. Although less commonly used, 'distribution free' is a name with much to commend it, because these tests ignore information on the pattern or scatter of observations which reflect the *distribution* displayed by the population. Technically the statistician uses the term *distribution* to describe in a particular way the underlying probability rules that generate the observations. The name *non-parametric* is used because distributions have their form determined by certain fixed quantities called *parameters*.

There are many different distributions and later in this chapter we shall meet the very important, but in some ways badly named, *normal distribution*. The name is unfortunate because it implies that other distributions are in some way abnormal. This is not so.

Non-parametric tests have the attraction that few assumptions are required for their validity; some, like the sign test, require little computation. They seldom make full use of the information in the data, although some of them use more than our friend the sign test.

Using the sign test with 12 observations, we get significance at the 5 per cent level only if 10 or more differences have the same sign. Intuitively we feel that, say, 4 negative signs of small numerical value and 8 positive signs of somewhat larger numerical value, might justify rejecting a hypothesis of zero mean population difference.

It is often possible to arrive at this conclusion using a *parametric* or *distribution-dependent* test. These tests also pave the way to statements about the likely magnitude of the population mean difference between, say, the two visual fields, although it is sometimes possible, but messy, to get these using non-parametric methods.

We chose the words 'quick and dirty' to describe the sign test; it is quick in requiring little computation, but dirty because it pushes a lot under the carpet, effectively replacing observed values merely by their sign.

A parametric test

Two key ideas we use are a *sample* mean and a *population* mean. If we have n observations in a sample we can denote their values in algebraic notation by $x_1, x_2, \ldots, x_n$. If the observations are the differences (LVF − RVF) given in column 4 of Table 9, then $x_1 = 7$, $x_2 = 16$, up to $x_{12} = 22$. If we want to refer to any typical sample value we may use the symbol x_i, where i (a *subscript*) denotes any number between 1 and n. The mean of the sample is usually denoted by $\bar{x}$ (pronounced 'x bar') or sometimes by m. It is simply the arithmetic mean or average of the sample values. We calculate it by adding together all the observations and dividing by the number of observations. In symbols

$$\bar{x} = \frac{x_1 + x_2 + \ldots + x_n}{n}.$$

Addition is such a common algebraic operation that mathematicians have invented a shorthand for adding a number of quantities for which a typical one is x_i. They write the sum as Σx_i. The symbol Σ (pronounced 'sigma') is the Greek equivalent of S and we may think of it as standing for 'sum'. Thus Σx_i is shorthand for

$$x_1 + x_2 + \ldots + x_n$$

and we may write

$$\bar{x} = \frac{\Sigma x_i}{n}.$$

In the visual response experiment our *population* mean refers to the true but *unknown* average value of (LVF − RVF) in the population. We usually denote this by the Greek letter μ (pronounced 'mew'). In this choice we follow a widely used convention in statistics of using Greek letters to denote constants characteristic of the population; these are the parameters we mentioned in the last section. It is important to distinguish the population mean (often, as in this case, unknown) from the sample mean (which can easily be calculated from the given observations).

Perhaps the most common kind of problem in statistics is that in which we have a set of observations (a sample) from which we want to infer something about the *unknown* mean of the population from which the sample has been taken. Let us now explore this problem systematically.

The hypothesis that there is no difference between the mean speeds of perception for LVF and RVF is essentially the hypothesis $\mu = 0$. This is often called a *null* hypothesis, a specialized use of the term. Many writers would call the hypothesis we considered in Chapter 2, that the doctor was guessing, a *null hypothesis* – using null in the sense that there was 'nothing' in the doctor's claim.

We can hypothesize values of μ other than zero. We shall formulate our theory to make possible also inferences involving other values. Whatever value we hypothesize for a parameter, we shall clearly be better disposed towards accepting it as the true value if the sample evidence supports it.

While the true population mean may be unknown it is nevertheless *fixed*. On the other hand the sample mean, $\bar{x}$, which we do know for our particular sample, is fixed only for that particular set of observations. Another set of observations (a new group of subjects in our visual response experiment) would give a different value for the sample mean $\bar{x}$. However, using random samples of any given size we may reasonably expect the values of $\bar{x}$ for different samples to cluster round μ in some fairly well-defined pattern. For some samples $\bar{x}$ will be less than μ, for others it will be greater. On the whole values far removed from μ will be less common than those close to it.

We may reasonably expect also that the means of large samples will be nearer to μ than those of small samples. For a given sample size, intuition again suggests that the more observations differ from one another the more $\bar{x}$ is likely to vary from sample to sample. These intuitive ideas about $\bar{x}$ should be reflected in inferences we make.

We have not yet said how we are to measure variation between observations; clearly we want some measure of their spread. A very simple measure of this is the difference between the largest and smallest observation (the *range*); it takes only the two extreme observations into account. In practice we prefer to take all observations into account.

Our usual choice is tedious to calculate but enables us to use readily available tables for our hypothesis tests. We calculate first the *estimated population variance*. It is formed by taking the squares of the deviations of each observation from the sample mean $\bar{x}$. These squares are added. By dividing their sum by n we get an *average* of these squared deviations; however, for technical reasons which we explain on page 81 we divide not by n, but by $(n - 1)$.

All this sounds rather a mouthful. Using our summation symbol Σ it can be written for a sample of n observations $x_1, x_2, \ldots, x_n$ with mean $\bar{x}$ as

$$s^2 = \frac{\Sigma(x_i - \bar{x})^2}{n - 1}.$$

In this formula $(x_i - \bar{x})$ represents the deviation of a typical observation x_i from the sample mean $\bar{x}$ and the summation is over the *squares* of these deviations for all n observations. The symbol s^2 is very commonly used for this estimate of population variance. Its positive square root s is the *estimated standard deviation*.

The formula we have just given for s^2 is algebraically respectable and demonstrates the logic of using s^2 as a measure of spread, but for computational purposes we nearly always manipulate it into a form that makes the calculation easier. We first square each observation and then add up the squares. Let's call their sum A. We next take the sum of all the observations, square this sum, and divide the result by the number of observations. Call this B. Then

$$s^2 = \frac{A - B}{n - 1},$$

or putting this all into algebraic symbolism in one bite it may be written

$$s^2 = \frac{\Sigma x_i^2 - \frac{(\Sigma x_i)^2}{n}}{n - 1}.$$

If you enjoy algebraic manipulations you might like to verify that the two formulae we have given for s^2 really are equivalent; don't worry if this is difficult – most people need practice to get this sort of manipulation right. If you don't feel like checking the equivalence algebraically try it for one or two simple numerical examples, e.g. 2, 5, 5, 6, 6, 7, 11. If you do the arithmetic correctly you'll find that each formula gives the answer $7\frac{1}{3}$. This example doesn't show, however, why in general the second formula is better for calculation. You'll see this more clearly if you take 2, 5, 5, 6, 6, 8, 11 as the observations, i.e. when the mean is no longer a whole number or integer.

Having obtained s we can easily calculate another quantity by dividing s by the square root of n, the number of observations; this quantity $\frac{s}{\sqrt{n}}$ is the *estimated standard error*, sometimes called just the *standard error*, of the mean.

To test whether a sample mean $\bar{x}$ 'supports' a hypothetical value μ for a population mean it is intuitively reasonable to look at the magnitude of the difference between $\bar{x}$ and μ. We write this $|\bar{x} - \mu|$. The vertical rules are called *modulus* signs; magnitude is never negative. Now $\bar{x}$ will vary more from sample to sample when variability between observations is high, so it seems reasonable to express $|\bar{x} - \mu|$ as a fraction of some measure of variation. The one that is chosen is the standard error of the mean and we calculate a statistic usually denoted by t, i.e.

$$t = \frac{|\bar{x} - \mu|}{\dfrac{s}{\sqrt{n}}}.$$

Large values of t cast doubt upon the hypothesis that μ is the true population mean. Before we decide what values of t will lead to rejection of our hypothesis, let's find the value of t for the differences (LVF $-$ RVF) in Table 9.

A spot of arithmetic

Battle stations now for an arithmetic sortie. Column 4 of Table 9 (page 68) gives the individual differences (LVF $-$ RVF) and provides our 'observations' for testing the hypothesis that the population mean difference μ is zero.

In the table we also have the total

$$7 + 16 + \ldots + (-1) + 19 + 22 = 198,$$

and the mean

$$\bar{x} = 198/12 = 16.5.$$

The sum of the squares of the entries in column 4 is 4396, i.e.

$$(7)^2 + (16)^2 + \ldots + (-1)^2 + (19)^2 + (22)^2 = 4396.$$

A pocket calculator is useful to work this out, especially if it has a memory store into which one can add each square as it is formed to the sum of the squares already obtained.

The square of the sum of the observations, divided by the number of observations is

$$\frac{(198)^2}{12} = 3267.$$

Thus

$$s^2 = \frac{4396 - 3267}{11} = 102.64$$

and

$$\frac{s^2}{n} = \frac{102.64}{12} = 8.55$$

and taking the square root gives the standard error

$$\frac{s}{\sqrt{n}} = 2.92.$$

To test whether the data are consistent with any hypothetical value of μ we consider

$$t = \frac{|16.5 - \mu|}{2.92}$$

where μ stands for any hypothetical value of the population mean that we may wish to test. In particular, to test $\mu = 0$ we have

$$t = \frac{16.5}{2.92} = 5.65.$$

To carry out the test we must decide whether this t-value of 5.65 falls into an appropriate critical region. As we have indicated, large values of t cast doubt upon the acceptability of our hypothesized value of μ. To determine the precise value required for significance at, say, the 5 per cent level requires some quite heavy computation; as for the chi-squared test we usually resort to tables to decide whether or not a t-value is significant. The tables we use, like those for chi-squared, depend upon *degrees of freedom*. In the examples in Chapter 3 we assumed fixed marginal totals in order to calculate expected values. We obtained degrees of freedom by considering the number of cells we could fill arbitrarily and still keep the correct marginal totals in the table.

In the t-statistic the analogue of an expected value in the chi-squared statistic is the fixing of $\bar{x}$ which appears explicitly in the numerator. It is also involved in determining s^2 since this is based upon deviations from $\bar{x}$. Now if $\bar{x}$ is fixed the total is fixed so that, if we know the values of $(n-1)$ of the observations we can work out the remaining one if we are to get the correct total simply by subtraction. Thus we assign $(n-1)$ degrees of freedom to t. This is not the only way of looking at degrees of freedom and in Chapter 5 we shall think about them in a more general context. The divisor $(n-1)$ used on page 77 in the formula for s^2 represents degrees of freedom.

Table 11 gives a table of t-values analogous to those for chi-squared in Table 5. As in that table the values are the minimum ones for significance at the levels stated at the top of each column, and for the degrees of freedom indicated at the left of the row.

Table 11 *Minimum t-value for significance at indicated level: two-tail test*

	Significance level %		
	5	1	0.1
Degrees of Freedom			
1	12.71	63.66	636.6
4	2.78	4.60	8.61
5	2.57	4.03	6.87
6	2.45	3.71	5.96
8	2.31	3.36	5.04
10	2.23	3.17	4.59
11	2.21	3.11	4.45
12	2.18	3.05	4.32
20	2.09	2.85	3.85
30	2.04	2.75	3.65
120	1.98	2.62	3.37

The values in Table 11 are for the *two-tail test*. This test is appropriate when we reject the hypothetical value of μ regardless of whether the evidence from the sample suggests a true value greater or smaller than the one we have been testing. Different tables are appropriate if we know, for example, that if the population mean is not μ then it must be greater than μ. A test called the *one-tail test* is then required.

Why do we use the expression 'tail'? Suppose that our hypothetical value of μ is the true population value. If we take lots of samples of a given size (and remember the sample size determines the degrees of freedom) we find values of t near zero occur more frequently than large values. Since we have taken the modulus or magnitude of t the value is always positive. Suppose now we consider a statistic t_0 that is the same as t except for the absence of vertical bars representing the modulus, i.e.

$$t_0 = \frac{\bar{x} - \mu}{\frac{s}{\sqrt{n}}}.$$

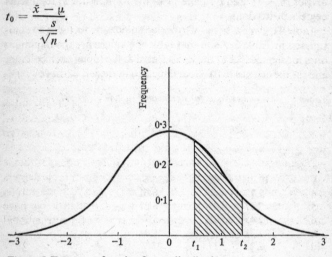

Figure 5 Frequency function for a t-distribution

A complicated formula enables us to work out the probability that t_0 takes a value in any given range *when μ is the true population mean*. Although the formula is complicated it is not difficult to show how it works graphically. The mathematical formula involves a *frequency function*. Figure 5 shows the values of this function on the vertical axis plotted against t_0 on the horizontal axis. The probability that t_0 has a value between, say, t_1 and t_2 is given by the shaded area in the figure. This area is often referred to as 'the area under the curve between t_1 and t_2'. For any specified number of degrees of freedom the frequency function has the same general 'bell' shape. It becomes flatter and more elongated the fewer the degrees of freedom, and more compact and peaked the higher the degrees of freedom. The frequency function has its maximum when $t_0 = 0$ or $\bar{x} = \mu$.

Now it's intuitively reasonable to take as our critical region values of t_0 with low probability when the hypothetical value μ is the *true* population mean but which would have a higher probability if the *true* mean had some other value. If we accept that the true value may be *any* other value, then both extremely large positive and extremely large negative values of t_0 tell against the hypothesis. These are the *tail* values. Because the frequency curve is symmetrical it is reasonable to take as our critical region sets of values of t_0 such that equal contributions to the critical region come from each tail – hence the expression 'two-tail test'. Thus for a critical region of size 0.05, corresponding to a 5 per cent significance level, we select a value of t_0 such that each tail contributes a probability of 0.025, i.e. one half of 0.05. Figure 6 illustrates the relevant tail areas. The values t_1 and $-t_1$ are selected so that the shaded areas each represent probabilities of 0.025.

Because we take equal areas from each tail, a value of t_0 that is either greater than t_1 or less than $-t_1$ for the appropriate degrees of freedom is equivalent to a value greater than t_1 in magnitude or modulus. Since the magnitude of t_0 is the statistic we have called t we may ascertain the critical values t_1 for the conventional

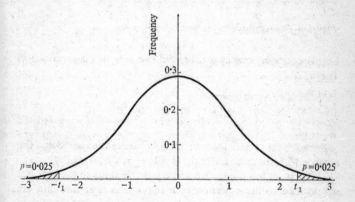

Figure 6 Critical region for a *t*-test

significance levels for various degrees of freedom from Table 11. For example, with 8 degrees of freedom, $t_1 = 2.31$ for significance at the 5 per cent level and $t_1 = 3.36$ for significance at the 1 per cent level.

In our visual response experiment we have 11 degrees of freedom. Table 11 shows that the *t*-value of 5.65 obtained for our data exceeds all values in the row corresponding to 11 degrees of freedom, viz. 2.21, 3.11 and 4.45. Thus we are justified in rejecting the hypothesis $\mu = 0$ at the 0.1 per cent significance level, i.e. the result is *very highly significant*.

Where has this got us? The numerical information that we threw away in the sign test has enabled us to establish significance more clearly – at the 0.1 per cent level instead of merely at the 1 per cent level. Here we must sound a note of caution. We shall learn in the next section that the *t*-test is not always so helpful. It requires a further assumption for its validity.

The *t*-test is sometimes referred to more fully as 'Student's' *t*-test, not because it has been used by a large population of students of statistics but because 'Student' was the pseudonym of its discoverer, W. S. Gossett (he was employed during the early years of this century by a well-known firm of Dublin brewers who encouraged their research workers to publish anonymously).

The t-test in trouble

Let us consider, one at a time, the two sets of observations in Table 12.

Table 12 *Two sets of observations*

| Set 1 | 17 | 26 | 30 | −2 | −1 | 30 | 44 | 30 | 20 | −10 | −5 | 19 |
| Set 2 | 1 | 2 | 3 | −1 | 2 | 3 | 1 | 2 | 2 | 1 | 2 | 180 |

For each set of data the total for all observations is 198 and the mean is 16.5, the same as for the (LVF − RVF) data.

The first set has 4 negative and 8 positive signs so the sign test would lead to non-rejection of a hypothesis that the data are consistent with a zero population mean. For this first set of data $t = 3.34$ (check this as an exercise to be sure you can calculate *t* correctly). The degrees of freedom are again 11, and from Table 11 we see that the result is significant at the 1 per cent level (critical value 3.11). Thus the *t*-test is again sharper than the sign test, a result we might have anticipated because of the additional information it uses on magnitudes of the observations.

Let's look now at the second set of data in Table 12. Here there is only one negative sign, so the sign test leads to rejection of the hypothesis of zero population mean at the 1 per cent level. The *t*-value turns out to be 1.11, well below the minimum critical value of 2.21 required for significance at the 5 per cent level with 11 degrees of freedom.

How can we explain the paradox? The sign test, which we know throws away a lot of information, indicates significance, whereas the *t*-test, which uses all the numerical information, does not.

Fortunately for the statistician's claim to follow common sense there is an explanation. The *t*-test has failed because it is not valid! The observation of 180 is the snag. The test has an important assumption built into it that we have not yet mentioned. To justify using a *t*-test we assume that our observations represent a sample from a *normal population*. This concept was described by an eminent statistician as 'too good not to be true'. It certainly is widely used. We associate with a normal population a bell-shaped frequency curve, not unlike that depicted in Figures 5 and 6 for the *t*-distribution frequency curve: the *normal distribution frequency curve*. A typical curve is shown in Figure 7.

The area under the curve between any two points on the horizontal axis, say x_1 and x_2, gives the probability that a sample value lies within the range between x_1 and x_2 when the population mean is μ; μ is always represented by the point on the horizontal scale immediately below the peak or highest point of the frequency curve.

The shape of the curve depends upon another factor that governs the spread. This is the *standard deviation* and is usually denoted by σ (pronounced sigma); this is the lower-case Greek letter equivalent to *s*. (We have already met the capital Σ used as a symbol for summation.) We recall that we named *s*, the positive square root of s^2, the *estimated standard deviation*; it provides

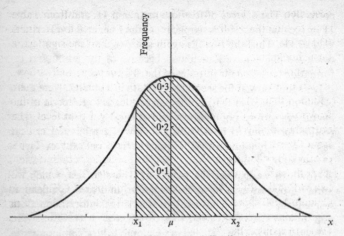

Figure 7 Frequency function for a normal distribution

quite a good estimate of the true standard deviation σ if our sample comes from a normal population.

We have not given the mathematical formula for the curve in Figure 7; it suffices to say that it is complicated for those unused to mathematical notation and would not be of much use even if we did quote it.

Sometimes there is justification for assuming our observations come from a normal population, but in most cases 'normality', as we usually call this, is assumed because it seems to work well in practice. But an unjustified assumption of normality can get us into trouble. This is what happened with the second set of data in Table 12.

We often escape such trouble because many of the tests we use are *robust*, that is, some of the assumptions which justify them in theory are not critical when it comes to applications. The *t*-test, in fact, has a reputation for robustness against departures from normality other than the extreme – and for the second set of data they are very extreme!

How do we tell whether it is reasonable to assume that our sample comes from a normal population? The experienced statistician usually casts his eye over the data and applies what is best described as his 'know-how'. There are formal statistical tests of data to indicate whether an assumption that they are a sample from a normal population is justified, but the following guidelines should help.

Generally normality implies that, in all but very small samples, small deviations of observations from the sample mean are more common than large ones. Although theoretical statisticians may frown upon this approach, we may get a good picture of what normality implies by comparing the magnitudes of all deviations from the sample mean with the *mean absolute deviation*. This is obtained by dividing the sum of all the magnitudes of deviations from the mean by the number of observations. An example will help to make this clear. In the data for (LVF − RVF) in Column 4 of Table 9 (page 68) the mean is 16.5 and the deviations are obtained by subtracting this from each observation. The values turn out to be

−9.5, −0.5, 13.5, −14.5, −0.5, 16.5, −2.5, 3.5, 3.5, −17.5, 2.5, 5.5.

The mean absolute deviation is obtained by adding the magnitudes of these deviations, ignoring the signs, and dividing this by 12; doing this gives a value of 7.5 in this case. As a rough rule of thumb the normality assumption is reasonable if about half, or perhaps a few more, of the deviations have a magnitude less than 7.5 and none are more than three or four times 7.5. Of the 12 deviations computed above, 7 are less than 7.5 in magnitude and the greatest deviation, 17.5, is less than three times 7.5.

Another indication of normality is that the numbers of positive and of negative deviations from the sample mean are about the same. There are 6 of each sign for the (LVF − RVF) data.

These general indicators of normality are intended as useful practical guidelines, not as hard-and-fast rules. Guides to normality may also be based on the value of s, but we shall not give details here, only point out that the guidelines are broadly similar to those using mean absolute deviation.

Now let us see how the data in Table 12 stand up to assumptions of normality. We leave the details as an exercise for the reader.

In the first set, 5 of the 12 observations give deviations that exceed the mean absolute deviation in magnitude; 3 more almost equal it, and the other criteria for normality look reasonably well satisfied also. The robustness of the *t*-test is sufficient to withstand any discrepancies.

But things are very different for the second set of data. All observations but one give negative deviations from the mean. Their values range from -13.5 to -17.5, while the sole positive deviation, associated with the observation 180, has the value 163.5! The *t*-test loses its theoretical basis in these circumstances, and we must turn to a non-parametric test. The only one we have described, the sign test, is not very efficient; better ones exist. A well-known one is the Wilcoxon test described in many text books including Campbell (1967).

Why is the normality assumption needed for the *t*-test? The reason is hidden away in the manner in which the critical values in the tables are calculated. This uses a mathematical formula devised on the assumption that our sample comes from a normal population.

The observation of 180 in the second data set gives cause for thought. It is so far removed from the values of the other data that it suggests an error. For obvious reasons such an observation is termed an *outlier*. Earlier in this chapter we spoke of making subjective judgements from time to time on such matters as the need for randomization. Judgements may also be needed in deciding what to do about outliers. If an observation is known to be wrong it should be rejected (or better still replaced by the correct value if ascertainable). If it is merely *thought* to be wrong subjective judgement is called for. This was involved with the decision in the visual field experiment to reject excessively long responses and to ignore missed responses. Given the safeguards already described this seemed reasonable in that experiment. The main difficulty facing the experimenters there was to decide what 'excessively long' was, a matter largely of practical 'know-how'.

An odd exceptionally long response implies a temporary lack of concentration and a missed one may have the same cause. The authors were wise to test for a significant difference in numbers of long or missed responses between the two fields, a difference which would in itself have implied different reaction patterns for the two fields.

Answering the right question

Testing whether or not we should accept the hypothesis $\mu = 0$ might be a useful preliminary step, but we can hardly regard it as more than that. The test has all the weaknesses of hypothesis tests that we talked about in Chapter 2 (pages 39–47). Also, even before he starts an experiment on visual responses a psychologist may well be convinced from his past experience that there will be *some* difference in mean response times between the two fields. Thus he would not expect the difference to be zero. He would be more interested in what the experiment can tell him about the actual *size* and *direction* of the difference.

In the paper from which we took the data of Table 9 there is a fascinating account of why we might expect differences and also a discussion of how these differences are affected both in magnitude and sign when recognition of digits is associated with 'secondary' tasks such as listening to music or speech at the same time.

Differences in response times between the two fields might only be of interest if they are on average above a certain threshold value, say, 20 milliseconds. These differences may be important in designing an instrument panel for some control system where warning signs call for urgent action by an operator. The panel should be designed so that devices requiring a quick response are in the visual field expected to give the more rapid recognition. Less important information sources can be relegated to positions where responses would be received in the other visual field. In some situations there might be good reasons to adopt this sort of layout for an instrument panel for a difference of 20 milliseconds but not for one of only 5 milliseconds.

One approach is to test not just the hypothesis $\mu = 0$, but a whole series with $\mu = 5$, $\mu = 20$, $\mu = 50$ and a host of other values in our expression for the t statistic. For some of these values we reject the hypothetical value of μ, while for a range of other values of μ we accept it. Fortunately we can formalize this 'hit-and-miss' approach and find in one operation all values of μ acceptable in a hypothesis test at a given level of significance.

We set about it this way. Let's write $t_{f,\alpha}$ for the t value that would just, and only just, give significance at the α per cent level with f degrees of freedom. From Table 11 we see, for example, that

$\alpha = 1$ and $f = 12$ give $t_{f,\alpha} = 3.05$,

whereas

$\alpha = 5$ and $f = 8$ give $t_{f,\alpha} = 2.31$.

Following from the discussion of critical regions in Chapter 2 (page 38) it becomes clear that if the μ we use in calculating our t statistic is the *true* population value then in $(100 - \alpha)$ per cent of a long run of experiments giving rise to a string of differing sample means, $\bar{x}$, the calculated value of t, i.e.

$$t = \frac{|\bar{x} - \mu|}{\dfrac{s}{\sqrt{n}}}$$

will be less than $t_{f,\alpha}$. Only in α per cent of our experiments in the long run would it exceed $t_{f,\alpha}$.

We now treat the expression

$$t_{f,\alpha} < \frac{|\bar{x} - \mu|}{\dfrac{s}{\sqrt{n}}}$$

as an algebraic inequality (the sign '$<$' means 'is less than') in which we know everything except μ. The rules for algebraic manipulation of inequalities lead us to the result that the inequality holds, providing μ lies between

$$\bar{x} - t_{i,\alpha} \frac{s}{\sqrt{n}} \text{ and } \bar{x} + t_{f,\alpha} \frac{s}{\sqrt{n}}$$

In other words, if the inequality holds for a particular value of μ we accept the hypothesis that μ is the population mean and the inequality will hold for any μ between

$$\bar{x} - t_{f,\alpha}\frac{s}{\sqrt{n}} \text{ and } \bar{x} + t_{f,\alpha}\frac{s}{\sqrt{n}}.$$

This defines a range of *acceptable* values of μ.

But we must look more closely at what this implies. We assumed above that everything was fixed, and that μ was unknown. This reflects the situation if we do *one* experiment. If we do another experiment of the same size, then $\bar{x}$ and s will be different and we will get a different range of 'acceptable' values of μ; but the 'real' μ is still the same fixed unknown constant.

Now it can be shown mathematically that there is a remarkable long-term reversibility in our argument. We have shown above that in any *one* experiment all acceptable hypotheses using a test at the α per cent significance level lie within an interval specified in terms of $\bar{x}$, s, $t_{f,\alpha}$ and n. The reversibility property is this: if we perform a large number of experiments and work out this acceptable interval each time, then in the long run $(100 - \alpha)$ per cent of of the intervals we work out will contain in them, or cover, the true *fixed unknown* population mean.

For this reason we call the interval we work out in the one experiment we actually perform the $(100 - \alpha)$ per cent *confidence interval* for μ. When we say that this interval contains μ we will be right in the long run for $(100 - \alpha)$ per cent of such statements. The '$(100 - \alpha)$ per cent' represents our *degree of confidence* that we are making a true statement, or our degree of belief that the statement 'the interval contains μ' is true.

The end points of the interval, namely

$$\bar{x} - t_{f,\alpha}\frac{s}{\sqrt{n}} \text{ and }$$

$$\bar{x} + t_{f,\alpha}\frac{s}{\sqrt{n}},$$

are the *confidence limits*.

For the visual field data for 95 per cent confidence limits we have $\alpha = 5$. Since $f = 11$, $\bar{x} = 16.5$ and

$\frac{s}{\sqrt{n}} = 2.92$, the 95 per cent confidence limits are

$$16.5 - (2.21 \times 2.92) = 10.05$$

and

$$16.5 + (2.21 \times 2.92) = 22.95.$$

The interval between these limits is the corresponding confidence interval.

We have seen that we can interpret the confidence interval in two ways: as the interval that contains all acceptable hypotheses using a test with critical region of size α; and in the sense that had we performed lots of experiments and calculated our confidence intervals each time, then $(100 - \alpha)$ per cent of our intervals would have spanned or included the unknown fixed μ.

To check the first interpretation the reader might like to verify that a value of μ from within the interval, say $\mu = 12$, gives $t < 2.21$ while a value from outside, say $\mu = 24$, gives $t > 2.21$.

The interpretation of confidence intervals and limits has been a source of controversy amongst theoretical statisticians since the 1930s. Some have not liked discussion of repeated experiments when only one has in fact been performed. Whatever the philosophical difficulties may be, the controversy has produced more heat than light. At the risk of offending theoretical statisticians – a rather easily offended breed – it is fair to assure the reader that if he understands what we have said about confidence limits he knows enough to interpret them sensibly in practice in contexts such as we are looking at here.

Generally speaking, if we reduce α and correspondingly increase $(100 - \alpha)$ we pay for the increased degree of confidence by the interval becoming longer. Again statistical analysis reflects common sense. The 99 per cent confidence limits for (LVF − RVF) population mean turn out to be

$$16.5 - (3.11 \times 2.92) = 7.42$$

and

$$16.5 + 3.11 \times 2.92 = 25.58.$$

Comparing these with the 95 per cent limits we see the obvious pay-off between increasing the length of the interval and increasing our confidence that the interval 'captures' the true population mean.

In this example note that neither the 95 nor 99 per cent interval includes zero. This is consistent with our rejection of the hypothesis $\mu = 0$. We see also how informative a confidence interval is compared to a hypothesis test. In the interval statement we learn something about *all* acceptable hypotheses and also get a good indication of a range of values highly likely to include the elusive unknown population mean.

If $\bar{x}$ is all we knew, it can be regarded as our most sensible 'guess' at the value of μ, and $\bar{x}$ is often called a *point estimate*. Our confidence interval provides an *interval estimate*.

In the examples we have discussed our point estimate $\bar{x}$ lies exactly at the mid-point of our interval estimate. This is often, but not always, the case in statistical practice; there are situations where the point estimate may be closer to one confidence limit than the other.

The *t*-test and related confidence interval statements can be generalized to more complex situations than have been considered in this chapter. However, their generality is still fairly limited and in this context the statistician could even do without the *t*-test (but *t*-tables are needed in other contexts, such as the applications in Chapter 5, pages 108–10). The situation is reminiscent of the song from *Annie Get Your Gun*: anything a *t*-test can do, the analysis of variance (which we meet in the next chapter) can do better. Nevertheless the *t*-test is widely used and enables us to introduce a lot of new ideas that we shall develop in the next chapter.

Some calculations to try

To be sure you've grasped the main ideas in this chapter try getting 99.9 per cent confidence limits for the mean drained weight of peaches in a certain brand of standard size cans. The weights in grams for a sample of seven cans are

282, 297, 283, 286, 291, 294, 283.

You may assume normality. Here's a tip to save yourself some arithmetic if you don't possess a pocket calculator. (Large

numbers won't frighten one of those, or you if you possess one.)
Subtracting 285 from all observations before we start calculations
has two effects: it reduces the sample mean by 285, and it gives
the same s^2 as we would have got from the original data. But look
how much simpler are the numbers we have to deal with, i.e.

$-3, 12, -2, 1, 6, 9, -2.$

The mean of these is $\frac{21}{7} = 3$. Adding 285 gives the mean of the
original data, 288. Now work out s^2 for yourself. You can also
check from the general formulae for $\bar{x}$ and s^2 that subtracting the
same number from all observations reduces $\bar{x}$ by that number but
leaves s^2 unaltered.

As usual, if you are stuck there are some notes to help in Chapter 13, pages 225–6.

5 The Analysis of Variance

Feeding the pigs

We warned that the last chapter would be heavy going and
sneaked in quite a number of new ideas. In this one we extend and
consolidate some of them but don't be disheartened if you find the
going is still tough. In the next chapter we take things easier.

Table 13 gives some data for the weight gained in one week by
pigs fed on three different diets. They were somewhat younger
pigs than the ones discussed in Chapter 1 and so put on less
weight. Eighteen pigs were used in the experiment, six at random
to each diet. We take the usual safety precaution of randomizing
to stop cheating in favouring one diet (even subconsciously) by
allocation of pigs that might be expected to do better for reasons
not connected with the diets.

Table 13 *Weight gains of pigs on three diets (grams)*

Diet							Mean
A	610	635	580	701	640	632	633
B	595	614	550	602	633	612	601
C	527	621	564	598	601	593	584
Grand mean (mean gain for all pigs)							606

The experimenter wants to know whether these results show a
real difference in response to the diets, and if one should be
recommended for general use. How can the statistician help?

We shall understand by the phrase 'real difference' a difference
in means for 'populations' corresponding to diets. A technique,
the *analysis of variance*, enables us to answer such questions, but
it is only valid given certain assumptions; its precise form depends
upon how the experiment has been conducted. Many textbooks
say quite a lot about analysis of experimental results but very

little about the design and conduct of experiments. *Experimental design* is an important part of the duties of an applied statistician; an excellent non-mathematical account is that by D. R. Cox (1958) in a book *The Planning of Experiments*. A more technical treatment of both the design and analysis of experiments is given by W. G. Cochran and G. M. Cox (1957).

We shall not deal in depth with design but merely give some indication of how the design of an experiment influences analysis and interpretation of results.

Experimental design is largely concerned with allocating units of experimental material to different treatments. Two important concepts are *randomization* and *replication,* both ideas we met in Chapter 4. Treatments, subject to certain restrictions mentioned below, should be allocated at random to units (as the diets for the pigs were). Replication, whereby we observe the effect of each treatment on more than one unit, provides a measure of 'noise' or sampling variation – the variation apart from that induced by the treatments we apply. We shall have more to say about restrictions on randomization in the section *Experimental Designs* (pages 105–8).

The data in Table 13 immediately show that the mean weight gain for pigs on diet A is higher than that for the other diets. The experimenter wants to know whether this superiority in the sample represents a real advantage for that diet in the *population* of all pigs of the type used in the experiment, or whether it simply reflects sampling variation. In other words, can the population mean for diet A, say μ_A, reasonably be supposed to be greater than the means for diets B and C, μ_B and μ_C? And he usually wants to estimate by how much the mean for one diet exceeds that for another in the population.

If the pigs were allocated to diets at random we can proceed with the formal calculations of an analysis of variance. As in the case of the *t*-test, we need other assumptions, including one of normality to perform hypothesis tests and to obtain interval estimates. We shall say more about these assumptions later.

Care is needed to ensure that the pigs are effectively allocated randomly to diets. It is permissible to use pseudo-random allocation if this means 'effectively' random. Experimenters sometimes think they are using pseudo-random methods when they are not. For example, the experimenter might have all his eighteen pigs in one pen and let them pass through a gate one at a time in any order *they* please, allocating the first six to diet A, the next six to diet B and the last six to diet C. The weakness of this method is that aggressive pigs are likely to pass through first; these might grow fastest irrespective of which diet they are given.

Truly random selection ensures that if there is no difference in the population means for the diets we shall reject the no-difference hypothesis only in the proportion of times implied by the size of the critical region. If the size of the critical region is $\frac{1}{20}$, we shall reject once in twenty tries in the long run.

Randomization is a relatively modern concept introduced by R. A. Fisher only a little over fifty years ago.

Basic ideas in the analysis of variance

The fundamental idea behind the analysis of variance is to partition the sum of squares of deviations of each observation from the mean into components that can be attributed to different sources of variation. In the simplest situation there are only two sources, but often there are more.

For the pig data the obvious sources of variation are differences between the average effect of diets, and differences between individual pigs fed on the same diet. The former is the systematic or imposed variation, the latter the sampling variation, noise or error.

The sums of squares of deviations that we shall have to calculate are reminiscent of the calculations of s^2 in Chapter 4 (pages 77–8). We gave two formulae for doing this, one being more useful in practice. In the analysis of variance we use analogues of these. We shall at this stage use the formulae that best indicate how the process works. We shall discuss more efficient computation in the next section.

We are faced with some arithmetic now which is easiest to follow with pencil and paper, so have these ready.

We have to talk about several different means. We refer to the mean of all observations as the *grand mean*. It is 606 for the data in Table 13.

Table 14 gives the deviations of each individual weight gain from the grand mean. The first pig receiving diet A, for example, which showed a weight gain of 610 g, was 4 g above the grand mean. On the other hand the third pig on diet B was 56 g below average. The sum of all the entries in Table 14 is zero – characteristic of the sum of a set of deviations from the mean, and provid-

Table 14 *Individual pig deviations from grand mean weight gain of 606 g*

Diet

A	4	29	−26	95	34	26
B	−11	8	−56	−4	27	6
C	−79	15	−42	−8	−5	−13

ing a useful check on the accuracy of the table. The sum of the squares of the deviations from the mean is

$$(4)^2 + (29)^2 + (-26)^2 + \ldots + (-8)^2 + (-5)^2 + (-13)^2$$
$$= 24\,980.$$

This is called the *total sum of squares of deviations from the mean*, a mouthful often abbreviated to the *sum of squares about the mean* or sometimes to the *total sum of squares*. This last wording is confusing because some writers use the phrase *total sum of squares* for the squares of the original observations (those in Table 13).

Let's consider now an artificial situation in which there are differences in responses by each group of pigs to the different diets, but no differences between responses of the individual pigs to any one diet. Suppose the response of each pig on diet A, then, is identical and the same as the mean response for that diet; the same applies to the other diets. Instead of the responses in Table 13 the individual responses are those in Table 15.

Table 15 *Weight gains of pigs if diet the only factor*

Diet

A	633	633	633	633	633	633
B	601	601	601	601	601	601
C	584	584	584	584	584	584

We easily verify that the grand mean of all observations in this table is again 606. Table 16 gives the deviations from the grand mean for the entries in Table 15.

Table 16 *Deviations of Table 15 'data' from grand mean*

Diet

A	27	27	27	27	27	27
B	−5	−5	−5	−5	−5	−5
C	−22	−22	−22	−22	−22	−22

Again the sum of all the deviations in Table 16 is zero. Also note that the means of the rows in Table 14 are 27, −5 and −22, the same as the corresponding row entries in Table 16. The sum of the squares of the deviations in Table 16 turns out to be 7428. This is described as the *sum of squares of deviations of the treatment means from the grand mean*, often abbreviated to the *between treatments sum of squares*.

If we compare Table 16 with Table 14 we see that the difference between two corresponding entries represents the deviation of the individual pig's weight from the mean for all pigs receiving the same diet; comparing Tables 15 and 13 gives the same result. This is our error or noise component of variation. These deviations are given in Table 17. As they are deviations from the individual diet

Table 17 *Individual pig deviations from diet means*

Diet

A	−23	2	−53	68	7	−1
B	−6	13	−51	1	32	11
C	−57	37	−20	14	17	9

means, the sum of the deviations from each diet mean is now zero. The sum of the squares of these deviations is 17 552. It helps to have a pocket calculator to do this sum – but it can be done on the back of an envelope! This is called the *sum of squares of deviations within treatments* or more briefly the *within treatments sum of squares*; other names widely used are the *error sum of squares* or the *residual sum of squares*.

Now here's a very important point. Adding the sum of squares between treatments of 7428 to the within treatments sum of squares we just obtained, 17 552, gives a total of 24 980, which was the sum of squares of deviations from the grand mean. This is no fluke, but follows from a relationship that may be established algebraically.

Our next task is to see how we use these quantities for hypothesis-testing and interval-estimation. To set us on the right tracks let's recall how we calculated s^2 when we did the t-test. We divided the sum of squares of deviations from the mean by the *degrees of freedom*, which in that case was 1 less than the number of observations because we had *fixed* the mean to calculate s^2.

Using the other way of looking at degrees of freedom (page 56) we may regard each observation as having a degree of freedom associated with it initially, but each time we 'fix' the value of something for a calculation (in this case the mean) we use 1 degree of freedom. If we have N observations altogether, we fix the grand mean to get the total sum of squares of deviations from it and thus this sum of squares has $(N - 1)$ degrees of freedom. For the pig experiment $N = 18$ and the total sum of squares of deviations therefore has 17 degrees of freedom.

What about degrees of freedom for the between treatments sum of squares? Here Tables 15 and 16 provide the clue. We have fixed all but 3 different numbers that appear in either of these tables, so there cannot be more than 3 degrees of freedom. Indeed, there are not even 3 because the values in Table 15 have been further constrained in finding the grand mean of 606 for all data, and those of Table 16 to give a zero total. Thus there are only 2 degrees of freedom for calculating this sum of squares of deviations.

There still remains the question of the number of degrees of freedom for the sum of squares within treatments. The 6 deviations from the fixed mean for diet A have 5 degrees of freedom. Similarly diets B and C each contribute 5 degrees of freedom to the sum of squares within treatments, making a total of 15. Since there are 2 degrees of freedom between treatments, the degrees of freedom between and within treatments add to the total degrees of freedom, i.e.

$$2 + 15 = 17.$$

This is the same result as we had with the sums of squares.

When we divide each sum of squares of deviations by its

degrees of freedom we get the *mean square* which provides us with an estimate of the variability attributable to a particular source. If the population means of the three diets do not differ, the mean square deviation within diets and the mean square deviation between diets should have much the same value; due to sampling variation they almost never have exactly the same value.

However, if the population means differ for the diets then the mean square deviation between treatments (i.e. the diets) will generally be larger than that within treatments.

Without going into the tedious algebra needed to establish this we can see it is intuitively reasonable: genuine differences between diets must tend to increase the sum of squares of deviations between diets above the value when there are no such differences, for in the latter case the value would only reflect variation due to 'noise'. If there are no differences between population means, the between treatment and within treatment mean squares both estimate the same quantity, called a *variance*. This variance is simply a measure of the natural (or noise, or sampling) variability between, in our case, individual pigs.

The practical problem is in deciding *how much bigger* the between treatments mean square must be before we can say the treatment means (diet means in the pig example) differ significantly. Everything goes smoothly providing we assume that our observations are all normally distributed, and have essentially the same 'spread' about the diet mean, the spread being measured by the variance. The mean square within treatments is in our case an estimate of this variance irrespective of any treatment differences. To test the hypothesis that the population means are the same for all three diets, or more generally that all of a set of population treatment means are equal, we calculate the *variance ratio* – abbreviated to V.R.,

$$\text{V.R.} = \frac{\text{Between treatments mean square}}{\text{Within treatments mean square}}.$$

It is customary to set out these results in *an analysis of variance* table. Some people use the abbreviation ANOVA for analysis of variance. Table 18 is the analysis of variance for the pig-feeding experiment results. The mean squares (column 3) are obtained by

Table 18 *Analysis of variance of pig diet data*

	Degrees of freedom (D.F.)	Sum of squares (S.S.)	Mean squares (M.S.)	Variance ratio (V.R.)
Between diets	2	7 428	3714	3.17
Within diets (error)	15	17 552	1170.13	
Total	17	24 980		

dividing each sum of squares by its associated degrees of freedom. The variance ratio of 3.17 is obtained by dividing 3714 by 1170.13.

We have already indicated that values of V.R. near 1 suggest consistency with the hypothesis of identical population means for the three diets, while high values suggest different means. If we can make the assumption of normality already discussed, we may use tables to decide whether a V.R. is significant. The test we apply is *robust* in the sense described in Chapter 4 – its performance does not depend critically upon the theoretical assumptions which justify it being fulfilled in the real data. We have to be more careful about these assumptions for interval estimation (discussed later in this chapter).

The tables in which we look up the minimum values for significance at a specified level are usually called *F*-tables in honour of R. A. Fisher, who first introduced the analysis of variance. Fisher originally proposed tables of a less direct kind. By necessity the tables are more extensive than the *t*-tables because we need separate entries for various combinations of two sets of degrees of freedom. Any reasonably comprehensive *F*-table occupies at least one page for each significance level. Table 19 is a very abridged table at the 5 per cent significance level covering only a small range of degrees of freedom.

Table 19 *Abridged F-table: minimum values of the variance ratio for significance at the 5 per cent level*

Numerator D.F.	1	2	3	4	6	10
Denominator D.F.						
5	6.61	5.79	5.41	5.19	4.95	4.74
8	5.32	4.46	4.07	3.84	3.58	3.35
10	4.96	4.10	3.71	3.48	3.22	2.98
12	4.75	3.89	3.49	3.26	3.00	2.75
15	4.54	3.68	3.29	3.06	2.79	2.54
20	4.35	3.49	3.10	2.87	2.60	2.35

Table 19 shows that there is a general decrease in the required critical value as the number of degrees of freedom increases (by moving from left to right or from top to bottom of the table). This is in line with common sense, for it tells us that in a bigger experiment a real difference between treatment means is indicated by a smaller excess over 1 of the variance ratio. So the greater the evidence leading to a given variance ratio the more confidently can we determine its message.

Once more we usually stick to conventional significance levels so that we can use standard tables; a well-programmed computer allows us to work out quickly the probability that any observed value of a variance ratio will be exceeded when the null hypothesis of no treatment difference holds.

From Table 19 we see immediately that, with the 2 and 15 degrees of freedom we had in the pig experiment, the minimum value of the variance ratio for significance at the 5 per cent level is 3.68. The value of 3.17 for V.R. in Table 18 is below this, and gives us no good reason for rejecting the hypothesis that all population diet means are equal. Remember our usual warning about hypothesis tests; this does not prove the hypothesis of equal means to be true – a larger experiment might well produce a significant result.

Indeed, there are few practical situations where we really believe there is no difference at all between population means for different treatments. There is almost always *some* difference and a non-significant result really implies only that the difference is too small to be detected by our experiment. Some people therefore rightly look upon an *F*-test (the usual name for the variance ratio test) as inappropriate except to indicate whether or not our experiment is large enough to detect small differences. When we are looking for treatment differences that may be of scientific or commercial importance we want our experiments to be large enough to detect these – if they exist.

In this pig-feeding experiment we are unable to reject the null hypothesis because of the marked variation in growth rates between pigs fed on the same diet – although pigs on diet A *seem* to do a little better. We can compare the situation with that in Chapter 3 where we drew labelled tickets from a hat. Suppose the hat held 18 labels each bearing the weight gain for a pig. If we draw the labels from the hat in three groups (equivalent to the three diets) of 6, it is unlikely that the 6 highest weights will all fall into one group. This *could* happen, but its unlikeliness make us favour an alternative hypothesis of a difference between diets.

It is possible (and also hard work) to work out the probabilities associated with all possible sample means of groups of 6 – those that give large differences between the means are less probable than those for which the differences 'average out'. Exact non-parametric tests for significance can be worked out along these lines. Tests based on the *F*-tables approximate closely to these and are completely valid when the required assumptions of normality hold.

The computations in practice

Most text books set about the calculations needed for an analysis of variance in a rather different manner to the one we have adopted. They use formulae analogous to the second one we gave for calculating s^2 in Chapter 4 (page 78). For example, for the total sum of squares of deviations from the grand mean the rule for calculating is

sum of squares of all observations
$$- \frac{\text{square of sum of all observations}}{\text{number of observations}}.$$

Note that at this stage we do not divide by the number of degrees of freedom as we did when calculating s^2. It is left as an exercise to the reader to verify that this gives the correct total sum of squares for the pig-feeding experiment.

Usually results are not quite as neat as they are for the pig data which we have fudged to give a nice integer mean rather than a nasty decimal expression. Thus had the mean for all the data been 606.055, instead of 606 exactly, the method of calculation just suggested would have been arithmetically superior to that used in the last section. If one has no pocket calculator, the calculations can be simplified with the device mentioned in Chapter 4 (page

94): subtracting 600 from all observations makes the arithmetic easier and does not alter the sum of squares of deviations. Try it!

A convenient way of calculating the between treatments sum of squares of deviations is this: first form the sum of squares of each treatment total divided by the number of units receiving that treatment; then subtract from this the square of the sum of all observations divided by the number of observations. That quantity is the same one we subtracted in getting the total sum of squares of deviations; it is often called the *correction factor*, because it adjusts a sum of squares to give a sum of squares of deviations from a mean.

The sum of squares *within* treatments, or error sum of squares, is now obtained by subtracting the sum of squares between treatments from the total sum of squares of deviations from the mean. The analysis of variance table can then be completed as before.

Experimental designs

The pig experiment had very little design, apart from giving each diet to the same number of pigs and allocating pigs at random to the diets. Sometimes it is appropriate to place restrictions upon the way we randomize. This in turn influences the analysis of results. The purpose of restrictions is to remove variation between individual experimental units that would otherwise appear in the within treatments or error sum of squares. We arrange things so that we compare treatments in more compact and homogeneous groups, referred to in statistical jargon as *blocks*.

A particularly simple case is the one in which each treatment is given to one and only one unit in each of several blocks, i.e. the number of units in the block is exactly equal to the number of treatments. This becomes a *randomized blocks* experimental design if we impose the additional rule that the treatments are assigned at random to the units within each block. To carry out significance tests using the *F*-tables we require the same normality assumptions as before and also an *additivity* assumption (explained later).

A well-designed experiment gives a smaller error mean square.

We shall illustrate the procedure using the pig data, making the further assumption that the three pigs in the first column of Table 13 all come from the same litter, all in the second column from another litter, and so on.

On biological grounds a group of pigs from the same litter should show less uncontrolled or random variability than a group of pigs from different litters. Genetic variability should be appreciably reduced by taking pigs from the same litter. Thus it seems intuitively reasonable to compare diets *within* each of the six litters. Analysis of variance provides a tool for doing just this as it removes from the error sum of squares differences between litters.

In general we speak of removing differences between blocks. We call these differences the *block effects*. Blocks should represent reasonably homogeneous groups of experimental material. For example, in a chemical experiment comparing three different methods of determining the calcium content of a rock sample we might employ five technicians and ask each to use each method once. Each technician would then constitute a block.

To determine the between blocks sum of squares in an analysis of variance we proceed in a manner similar to that for obtaining the treatment sum of squares in our earlier example. In effect we calculate what the sum of squares of deviations from the grand mean for all observations would have been if each entry were replaced by its block mean. When each litter is regarded as a block we find from Table 13 that the mean for the first litter, for example, is

$$\frac{(610 + 595 + 527)}{3} = 577\tfrac{1}{3}.$$

It is now possible to work out a between litters sum of squares by replacing individual observations by deviations of block means from the grand mean and squaring these. For the first block (litter) each deviation is

$$577\tfrac{1}{3} - 606 = -28\tfrac{2}{3},$$

and we may obtain the between litters sum of squares by squaring all the corresponding deviations. The result is a sum of squares of

11954. It has 5 degrees of freedom. (Why?) The reader should confirm the value of this sum of squares. Do this either directly or using a formula analogous to the one for the treatment sum of squares which we may write as

$$\frac{\text{sum of squares of block totals}}{\text{number of units in each block}} - \frac{\text{sum of squares of all observations}}{\text{number of observations}}.$$

Since each treatment occurs exactly once in each block, the number of units in a block is equal to the number of treatments. Some authors extend the definition of a randomized blocks design to include designs where each treatment occurs a specified but equal number of times in each block rather than just once. This requires relatively small modifications in the analysis.

The degrees of freedom and sums of squares for blocks are subtracted from the corresponding within treatments degrees of freedom and sums of squares in Table 18 to give new and reduced error degrees of freedom and sums of squares. If the population differences between diet means are zero, then the between treatments mean square should once again have a value about the same as the new error mean square. If the treatment mean square is appreciably greater than the error mean square a real difference between treatment means is indicated. Given the assumption of normality, significance is again determined using an F-table providing the *additivity* assumption holds. The additivity assumption essentially means that any applied treatment must have the same effect in each block. By 'has the same effect' we mean that if a treatment for instance boosts one observation by two units in one block it may be expected to give the same two-unit boost in any other block, apart from random variation.

The new analysis of variance taking litters as blocks is given in Table 20. The new error sum of squares is

$$17\ 552 - 11\ 954 = 5598$$

with 10 degrees of freedom.

The variance ratio for the difference between diets is increased from 3.17 in Table 18 to 6.64 in Table 20. From Table 19 we find

Table 20 *Pig data: randomized blocks analysis*

	Degrees of freedom	Sum of squares	Mean squares	Variance ratio
Between blocks	5	11 954		
Between diets	2	7 428	3714	6.64
Error	10	5 598	559.8	
Total	17	24 980		

that the value required for significance at the 5 per cent level with 2 and 10 degrees of freedom is 4.10. Thus the new analysis indicates significance whereas the original one did not. The reduction in denominator degrees of freedom has increased the F-value required for significance from 3.68 to 4.10, but this is more than offset by the reduction in the error mean square.

Statistically speaking, we have acted improperly in using the same data to illustrate two analyses leading to different conclusions; only one of these can be valid in a real situation. If the pigs *were* grouped by litters the analysis in Table 20 is correct; if they were *not*, the Table 18 analysis is appropriate. Both analyses assume that the appropriate randomization has been carried out and that the 'noise' component has a homogeneous normal distribution. The second analysis also assumes that effects of 'classifications', i.e. blocks and treatments, are additive.

There are considerably more complex designs than randomized blocks. Most have a clearly defined pattern but in some not every treatment appears in each block. Computation of the valid analysis becomes more difficult for sophisticated designs.

It is possible to remove other sources of variation by having further classifications or different modes of classification.

Interval estimates: being realistic

By now we know a good deal about the limitations of hypothesis tests. The pig experimenter wants to know what his experiment tells about the reality and magnitude of differences between diet means for the population. He may want a confidence interval for his estimate of the population mean for a particular diet. For diet A the 'sample' mean of 633 grams weight gain provides a point estimate of the population mean for that diet. To get confidence limits for this estimate we follow a very similar path to that in Chapter 4 (page 90). The s used in confidence limits on page 90 is replaced by the square root of the error mean square. The degrees of freedom f for $t_{f,\alpha}$ become those of the error mean square, not $(n - 1)$. The n (page 90) is replaced by the number of replicates, more usually denoted by r. Suppose $\bar{x}_i$ denotes the mean for all units receiving treatment i, then the $(100 - \alpha)$ per cent confidence limits for the corresponding population mean μ_i, say, are

$$\bar{x}_i - t_{f,\alpha}\frac{s}{\sqrt{r}} \quad \text{and} \quad \bar{x}_i + t_{f,\alpha}\frac{s}{\sqrt{r}}$$

where $t_{f,\alpha}$ is obtained from the appropriate t-table.

Applying this to diet A in Table 18 we have $f = 15$. We find the 95 per cent confidence limits using tables slightly more extensive than Table 11 to give $t_{f,\alpha} = 2.13$. (We could derive this as an approximation from Table 11 which shows critical values for 12 and 20 degrees of freedom of 2.18 and 2.09 respectively. The general pattern of the table indicates that the value for 15 degrees is roughly half way between these two values.)

Thus 95 per cent confidence limits for the population mean for diet A would be

$$633 \pm 2.13\sqrt{\frac{1170.13}{6}},$$

where the symbol $\pm$ means we add or subtract what follows to get the two limits. These are 603.3 and 662.7 so that the 95 per cent confidence interval for μ_A is 603.3 to 662.7. Let's recall what this means. The '95 per cent' reflects our degree of belief that the interval includes the true but unknown mean, 'belief' in the sense that in the long run 95 per cent of the intervals produced in a series of experiments would span the true mean.

We can, in a similar way, work out 95 per cent confidence limits for diets B and C from the Table 18 analysis; they turn out to be 571.3 to 630.7 for diet B and 554.3 to 613.7 for diet C. Note that for all three diets the length of the confidence interval (the difference between the upper and lower limit) is the same, 59.4.

Because it involves a smaller error mean square the analysis in

Table 20 gives shorter confidence intervals for the means. For diet A the limits are

$$633 \pm 2.23\sqrt{\frac{559.8}{6}}$$
$$= 633 \pm 21.5,$$

i.e. 611.5 and 654.5.

(Note the different value of $t_{f,\alpha}$. Why?) Limits for the other two diets may be obtained in a similar way, all the intervals now having a length of 43.0. A shorter interval for a randomized block experiment compared to a simpler experiment is common, and the more highly designed experiment is said to have higher *precision*. One aim in choosing an experimental design is to select a design of high precision in the sense of having a small error mean square, leading to shorter confidence intervals in general.

We may also calculate confidence limits and intervals for comparisons between treatments; in particular we may calculate confidence limits for the difference between two treatment means. When each treatment is replicated the same number of times, say, r, $(100 - \alpha)$ per cent confidence limits for the difference between corresponding population means are given by

$$\bar{x}_i - \bar{x}_j \pm t_{f,\alpha}\, s\sqrt{\frac{2}{r}},$$

where $\bar{x}_i$ and $\bar{x}_j$ are sample means for the two treatments of interest, the ith and jth treatment. Mathematical justification of this result requires some statistical theory, but we shall show in the next section that it gives the correct result when we re-examine, (from an analysis of variance viewpoint), the data of the previous chapter.

On the basis of the analysis in Table 18, the 95 per cent confidence limits for the difference between population means for diets A and B turn out to be

$$633 - 601 \pm 2.13\sqrt{\frac{2 \times 1170.13}{6}}$$
$$= 32 \pm 41.9,$$

i.e. −9.9 and 73.9.

Note that the difference includes zero so we would not reject the hypothesis that the population means were equal.

In the analysis of Table 20 the corresponding limits turn out to be 1.6 and 62.4. The interval no longer includes zero: consistent with the overall F-test used on the variance ratio in Table 20, we reject a hypothesis of no difference between diets.

Care is needed in using and interpreting confidence limits for differences between means when comparing more than two treatments. The overall F-test may indicate a difference between population means, yet confidence intervals for differences between some pairs of means may include zero. Also when the F-test does *not* indicate an overall difference between means one may find confidence intervals for differences between two means that do not include zero. Strictly speaking, the confidence limits just given are valid only for the difference between means of two treatments *selected at random*! It is unlikely anybody would do precisely this in real life. One must not conclude that the largest and smallest treatment means differ significantly simply because the confidence interval for their population difference does not include zero. With large numbers of treatments this may, and frequently does, happen even when an overall F-test indicates no significant difference between treatment means. To explain this paradox fully requires a working knowledge of *distribution theory*.

We can say from the practical viewpoint that if the confidence interval for the difference between two means *does* contain zero, it is safe to say that the population means for those two treatments do *not* differ significantly. However, the confidence intervals for some treatment differences may exclude zero without the differences really being significant. The problem is that of *multiple comparisons* and there is a large literature on the subject. The best advice to an experimenter in doubt is to consult a statistician. Even the statistician may be a little evasive on problems in this area because the appropriate action depends very much upon what assumptions one is prepared to make. There are still a good many loose ends in the theory.

One final note of caution about confidence limits. While hypothesis tests using variance ratios are robust against moderate departures from the assumptions of additivity and homogeneous 'normal' error structure, confidence intervals are not; if these assumptions do not hold for a given body of data, the intervals may not be valid. There are 'cures' for these problems, but the cure varies with the disease, so the best advice we can give is to consult a statistician or a text book that deals with the subject in depth, e.g. Cochran and Cox (1957).

Analysis of variance and t-tests

We could have analysed the visual response data in Chapter 4 by an analysis of variance instead of a *t*-test. More calculation is involved, but we shall carry it through to demonstrate the correspondence between the two methods. Instead of working with the difference (LVF $-$ RVF) in column 4 of Table 9 (page 68) we work with the data for the individual fields in columns 2 and 3 of the table. Here LVF and RVF are the 'treatments', the subjects numbered 1 to 12 the 'blocks'. Randomization within each block is not possible in the usual sense, because a person's left or right visual field is fixed. Effective randomization was attained in the experiment by presenting the stimuli in random order to one or the other field.

To carry out the analysis we need the 'block totals' (LVF $+$ RVF) for each subject. For the first subject this is 1121, for the second 1026, and for the twelfth 912. Using the formulae we have developed in this chapter for the various sums of squares we get for the total sum of squares

$$(564)^2 + (521)^2 + \ldots + (445)^2 - (12\,358)^2/24 = 42\,011.83.$$

Here 12 358 is the total of the observations for all subjects in both fields. The 'treatment totals' are the totals for LVF and for RVF over all subjects, i.e. 6278 and 6080, the totals of columns 2 and 3 in Table 9. The sum of squares for treatments is

$$\frac{(6278)^2 + (6080)^2}{12} - \frac{(12\,358)^2}{24} = 1633.50.$$

The sum of squares between blocks (subjects) is

$$\frac{(1121)^2 + (1026)^2 + \ldots + (912)^2}{2} - \frac{(12\ 358)^2}{24} = 39\ 813.83.$$

By subtraction we obtain the error sum of squares as

$$42\ 011.83 - 39\ 813.83 - 16\ 33.50 = 564.50.$$

Table 21 gives the appropriate analysis of variance.

Table 21 *Analysis of variance for visual field data*

	Degrees of freedom	Sum of squares	Mean squares	Variance ratio
Between subjects	11	39 813.83		
Between L VF and RVF	1	1 633.50	1633.50	31.83
Error	11	564.50	51.32	
Total	23	42 011.83		

The relationship to the analysis in Chapter 4 is not immediately obvious, but it is there. For example, the error sum of squares of 564.50 in Table 21 is exactly one half the sum of squares of deviations of the differences (LVF − RVF) (column 4 of Table 9) from their mean of 16.5. Also the sum of squares between LVF and RVF in Table 21 of 1633.50 is $6 \times (16.5)^2$; again it involves the mean difference of 16.5. Most important of all, the variance ratio of 31.83 is the square of the t-statistic 5.65 obtained in Chapter 4 (allowance again must be made for a slight round-off in computation).

There is a definite relationship between a t-statistic with f degrees of freedom and a corresponding variance ratio with 1 numerator and f denominator degrees of freedom. The latter is always the square of the former. This relationship is reflected in the t- and F-tables where the critical value at a given significance level in the F-table corresponding to 1 numerator and f denominator degrees of freedom is the square of the corresponding t-value with f degrees of freedom.

In Chapter 4 (pages 91–3) we obtained confidence limits for the difference (LVF − RVF). We may obtain the same limits using the formula given in the last section for the confidence limits for the difference between two treatment means. As the difference between the treatment means for LVF and RVF is 16.5, appropriate 95 per cent confidence limits are given by

$$16.5 \pm 2.21 \sqrt{\frac{2 \times 51.32}{12}}$$

and it is easily verified that these limits are those obtained in Chapter 4. Thus a particular numerical case justifies the expression for relevant confidence limits given in this chapter.

Because the precise form taken by an analysis of variance depends upon the experimental design, the reader who has to design and analyse experiments frequently should seek the aid of a statistician or use one of the recommended text books on experimental design.

Try one yourself

Now have a go at analysing these data. Three brands of petrol are used in a car mileage test. Four different types of car are involved, and there are three cars of each type. Each brand of petrol is randomly assigned to one car of each type. Do the data in Table 22 suggest that the petrols differ in average performance?

Table 22 *Performance of various types of car with differing brands of petrol (miles per gallon)*

	Car type			
Petrol brand	1	2	3	4
A	42	37	49	55
B	39	38	47	53
C	41	38	49	52

In this example different types of car correspond to 'blocks' and different brands of petrol to 'treatments'.

To make computation easier we suggest you subtract 40 from all observations before you start. This does not affect the overall *F*-test. How will it affect confidence limits for brand means or for differences between brand means? As usual, we give hints in Chapter 13 (pages 226–7) if you are stumped.

6 Interlude – Let's be Less Conventional

Numbers and literary style

The last couple of chapters have been heavy going. We've introduced many new ideas and it's time for something lighter. While you're getting back your breath let's look at two applications of statistical ideas that don't involve us in deep new principles.

On the surface the first problem we're going to look at doesn't seem numerical at all. We're going to look for differences between the literary styles of two authors. To make this amenable to statistical study we have to translate our information into numerical form.

As a simple example of how to study literary style statistically let us consider two of the presidential addresses to the Royal Statistical Society published in one of the Society's journals. One piece of numerical information we can extract from such an address is the number of words in each sentence. Interesting differences between authors are reflected in these numbers.

Differences in style also reflect themselves in the relative use authors make of long and short words and in the frequency with which they use certain words or phrases. Table 23 gives the lengths of each sentence on the first pages of the presidential addresses for 1972 and for 1973, and Table 24 the numbers of words of between two and five letters on those same two pages.

Table 23 *Sentence length: presidential addresses*

Date	Length of successive sentences (*words*)										
1972	34	8	12	47	22	35	26	27	42	32	18
1973	11	31	45	31	12	31	39	16	21	31	36
1972	62	6	57	10	57	7					
1973	28	31	39	31	22	33					

What can we glean from these data without formal analysis? Table 23 suggests that the 1972 president used sentences of more

Table 24 *Frequency of short words: presidential addresses*

| | Number of letters | | | |
	2	3	4	5
Date				
1972	86	91	53	47
1973	84	86	58	45

variable length than his 1973 counterpart; sentence lengths varied between 6 and 62 words in 1972 but only between 11 and 45 words in 1973. Also in 1972 six sentences were less than 20 words long and five sentences more than 40 words long; in 1973 only three sentences were less than 20 words and only one more than 40. These data look clearer in Table 25.

Table 25 *Numbers of short and long sentences: presidential addresses*

	1972	1973
Under 20 words	6	3
Over 40 words	5	1

The average lengths of sentences were very similar, 29.5 words per sentence in 1972 and 28.7 in 1973. Thus the main difference in sentence construction was in variation of length rather than in average length. Table 24 indicates little difference in the frequency of short words.

Does variation in sentence length imply a very different literary style? Although each president was talking about statistics they dealt with different aspects of the subject and their ways of expressing themselves could have been influenced by this. However, the opening page of their addresses contained general remarks and did not involve discussion of technical matters. It seems likely that differences in sentence lengths do reflect characteristics of the styles of the two presidents. Both authors were eminent members of the Royal Statistical Society, the first a man whose intelligent interest in statistics is secondary to his main occupation and the second an eminent contributor both to statistical theory and practice. The first was the Rt Hon. J. Harold Wilson, at that time Opposition leader in the British Parliament, and the second was the professor of statistics at Edinburgh University, D. J. Finney.

It is interesting to speculate if the differences in style may reflect the contrasting interests and training of the two men. Mr Wilson is a forceful political debater. It seems not unreasonable for him to use long sentences while developing an argument with an occasional short pithy sentence to drive home a point. On the other hand a person primarily concerned in his professional life with logical deductions intended to appeal to reason rather than emotion may prefer to compartmentalize each step of an argument into convenient portions. Either deliberately or subconsciously he may try to subdivide an argument into not too long sentences. He would have little reason to use short sentences for their dramatic impact.

Even in a relatively simple study like this some further points need noting. Firstly, a unit of one page is not perfect for a comparative count – even if we compare pages from the same journal. Indeed, in 1973 there were four fewer lines on the first page than in 1972 due to the inclusion of a sub-heading (excluded from the count). Secondly, in counting we must decide on such matters as whether hyphenated words count as one or two words. The 1972 address contained six hyphenated words on the first page and the 1973 address only one (this in itself might indicate a stylistic difference). Thirdly, should names of persons, places, etc., be omitted from the count or perhaps scored separately? Sometimes their occurrence follows from the topic and is independent of the author's style, or an author may frequently refer to people or places because he is a habitual 'name-dropper'.

We may carry the numerical study of linguistic style to more formal levels. There are, for example, statistical tests that help us to decide whether there is a real difference in the spreads of sentence lengths used by different authors – the sort of thing we suspected in our example – even though the average sentence lengths are about the same. A fascinating account of statistical analysis of literary styles may be found in Williams (1970).

Interpreting answers when you don't know the question

The second problem of this interlude chapter has come into prominence in recent years under the name of *randomized responses.* A session was devoted to this topic alone at the leading world conference of statisticians in 1975 – that of the International Statistical Institute in Warsaw. It provides a means of overcoming people's reluctance to give an honest answer to a question in a sensitive area involving such matters as moral problems or tax evasion. The worry people have about honestly answering questions in these areas is that the answers could be traced back to them.

Let's look at how the method works. Suppose we want to estimate the proportion of taxpayers who do not declare all their taxable income. However strong the guarantees of confidentiality there will be some reluctance to admit to tax dodging. It is hardly good enough to base an estimate of the proportion of tax dodgers on a random sample where people are simply asked point blank 'Did you dodge any tax last year?'

With the randomized response technique, which is ideal for questions having a 'yes' or 'no' answer, we proceed like this. We select a random sample of, say, 120 taxpayers. We then give each one a sheet of paper with these questions on it:

1. Is the last digit of your year of birth odd?
2. Did you declare all your taxable income on last year's return?

We next supply each person with a die (yes, that's the singular of dice) and ask him or her to cast it without disclosing the result to anybody. We then instruct those who cast a 5 or 6 to answer the first question, and those who cast a 1, 2, 3 or 4 to answer the second question. They do *not* disclose to anybody which question they answered.

If 44 of the 120 answers are 'yes', what can we say about the number of tax dodgers in the sample? The probability of getting a 5 or 6 when a die is cast is $\frac{1}{3}$ so approximately 40 people in the sample of 120 answer the first question. Now a person is almost equally likely to be born in a year with an even last digit or with an odd last digit (since fluctuations in birth rate are small); an equal probability assumption is reasonable. About half of those answering the first question put down 'yes'; about 20 'yes' answers come from question 1. Since 44 people in all answer 'yes', this gives $44 - 20 = 24$ as the expected number – or our best guess at the number – of people answering 'yes' to the tax evasion question. In all, the expected number of people in our sample of 120 answering the second question is $120 - 40 = 80$, so that we estimate the proportion of honest taxpayers in our sample at 24 out of 80, which is 30 per cent.

Sampling variation enters into our calculations in two ways. Firstly, different samples of 120 would contain different numbers of tax dodgers; but we may reasonably hope the proportion in a random sample reflects the population proportion, i.e. that the sampling error is not large. The other source of variation is in casting the die and in the proportion of 'yes' answers to question 1. Try casting a die yourself 120 times and see how many 5 and 6 you get. There should be about 40 – values like 37, 42, 39 are quite likely, but if you get numbers like 11 or 79 either you have a loaded die or you are an accomplished cheat! Of those who answer the first question it is unlikely that *exactly* half will answer 'yes' even if we assume they all answer the question honestly – at best we can hope that about one half will say 'yes'.

Allowing for all these sources of 'error' we can work out a confidence interval for the population proportion based upon our sample result. Let's leave that to the experts who can also advise on the best sort of proportions of people answering each question. Clearly this proportion depends on the 'random' choice mechanism; if we only ask those scoring a 6 with the die to answer question 1 the expected number answering it would be reduced to 20. It is a mixture of the art and science of statistics to choose the right proportion. Generally the fewer who answer the harmless question the better we can estimate the population response to the delicate question, providing we still get honest responses. On the other hand if participants know that few people are answering the harmless question there is a danger that people will become reluctant to answer the delicate question honestly.

What evidence there is suggests that randomized responses give more reliable results than direct questioning when people feel they may be answering an embarrassing or dangerous question. Wrong answers may be given to sensitive questions either because people lie to hide their misdeeds, or because they feel bold in falsely claiming to practice some vice. It is unlikely that the two sources of error cancel one another out in any one situation.

Some surveys of human sexual behaviour done by conventional methods were widely criticized by statisticians who felt there were dangers of this type. The temptation to give a misleading answer is likely to be reduced when nobody can possibly know what question is being answered.

Two key pieces of information we need to make our example work are a knowledge of the proportions answering each question and the expected proportion of 'yes' answers to the harmless question.

Variations on the theme

There are several amusing variations on the randomized response theme. One arises in the true-or-false response situation where either response could embarrass according to which question one is answering. For example, the questions may be the statements:

1. I submitted an incorrect tax return last year.
2. I submitted a correct tax return last year.

Suppose that, much as in the last example, we again cast a die and instruct people to write 'true' or 'false' in response to the first statement if they throw a 5 or 6, and to the second if they throw between 1 and 4. Suppose our sample is 126, and we want to estimate the proportion P in the population who have submitted false returns. Suppose 69 people give an answer 'true'.

We now argue this way. Incorrect returns are submitted by those who answer 'true' to the first statement or 'false' to the second. The answer 'true' to the first statement has probability P, since probability corresponds to proportion here. The answer 'true' to the second has probability $(1 - P)$, because the probability of an event plus the probability of an opposite event always equal 1 (page 35).

Now the probability of throwing 5 or 6, and so of replying to the first statement, is $\frac{2}{6}$ or $\frac{1}{3}$. The event that the response is 'true' is independent of the die-casting result. Thus we may use the multiplication rule and find the probability that statement 1 is the selected statement and the response is 'true' to be

$$\tfrac{1}{3} \times P.$$

We may call this the probability of a 'true' response to the first statement. Similarly the probability of a 'true' response to statement 2 is

$$\tfrac{2}{3} \times (1 - P)$$

since the die casting rule tells us that $\tfrac{2}{3}$ of the people respond to this question. 'True' responses to either statement provide the two mutually exclusive ways of getting a 'true' response. We know that 69 of 126 answered 'true' so the probability of a 'true' response is $\tfrac{69}{126}$. Using the addition rule for mutually exclusive events we can estimate P by solving the equation

$$\tfrac{69}{126} = \tfrac{1}{3}P + \tfrac{2}{3}(1 - P).$$

The solution of this equation is $P = \tfrac{5}{14}$. Thus our best estimate of the percentage submitting false returns would be

$$\tfrac{5}{14} \times 100 = 35.7.$$

Again techniques exist that enable us to put confidence limits on this estimate.

This community looks more honest than the last, where only 30 per cent appeared to submit correct returns.

The formulation we have just used for the problem has an obvious advantage over the earlier one in avoiding biased answers; clearly there is no advantage to an individual to give an incorrect answer to either statement. Under the first formulation some may have felt 'yes' was the only answer that could never be incriminating, and thus the safest answer irrespective of the question being asked. In the second example no stigma can be attached to either answer because 'false' is a socially acceptable response to the first statement while 'true' is socially acceptable for the second.

Further modifications are possible. In one we avoid needing to know the proportion expected to answer 'yes' to a harmless question; we give the question to two distinct samples using a randomizing device that ensures that *different* known proportions get the harmless question in each sample. We shall not go into details but typical pairs of questions might be:

1. Do you own a pet cat?
2. Do you take drugs illegally?

In yet a further modification people are asked to answer 'yes' or 'no' to no question at all. It works like this. A box of tickets contains exactly the number of tickets there are people in the sample. Each ticket has a number – either 1, 2 or 3 – on it and the total number of tickets bearing each of these numbers is known to the organizers of the survey. Each person is asked to draw a ticket without disclosing the number on it. The instructions are to write 'yes' if the ticket drawn bears the number 1, to answer the question 'Do you make private telephone calls on your office telephone without paying for them?' if the number is 2, and to write 'no' if the number is 3.

Suppose we give 100 tickets, 30 labelled 1, 40 labelled 2 and the remaining 30 labelled 3, to 100 people who give us 65 'yes' answers and 35 noes. 30 of the 'yeses' are from holders of tickets labelled 1, implying that 35 have answered yes to the question. Since 40 tickets are labelled 2 this implies that 35 out of 40 or 87.5 per cent make private calls on office phones without paying.

For the reader wanting further examples and more details there is a paper by Campbell and Joiner (1973) entitled 'How to get the answer without being sure you've asked the question'.

Keeping yourself amused

This chapter being an interlude, we won't try anything taxing, but if you want to work your way through a randomized response problem try this one. Suppose you ask 135 married men two questions:

1. Is your age divisible by three?
2. Do you indulge in extra-marital sexual activities?

As before, they all cast a die. If a 5 or 6 is scored they answer question 1, otherwise question 2. If 72 out of 135 answer 'yes', show that our estimate of the percentage who indulge in extra-marital sexual activities is 63.3 per cent. Hints as usual are in Chapter 13 (pages 227–8).

If you want to compare literary styles, contrast the distribution of sentence lengths on a couple of pages of this book (choose complete pages without tables or diagrams) with the distribution in some book of quite a different kind – perhaps an Agatha Christie thriller. Have a look at the frequency of occurrence of words of various lengths in the two works as well.

7 Relationships between Variables

Relationships – meaningful and otherwise

This chapter is about two widely used statistical ideas – *correlation* and *regression*. These are used when we look at association and at relationships, but many people have a tendency to read too much into them, especially into correlations. It's terribly easy to point to cause and effect on a purely statistical basis without any hard evidence.

The smoking and lung cancer controversy provides a good example. Statistics alone can never prove that cigarette smoking is a cause of lung cancer, but statistical evidence is a suggestive pointer.

Statistics indicate that there is a higher incidence of lung cancer among those who smoke many cigarettes than among those who smoke few or none at all. This suggests but does not prove a link.

Let's look for a moment at some other figures. Comparison of the records for the number of television licences taken out and the number of people treated for mental disorders over a period of years show a clear relationship – the more television licences taken out the more people receive psychiatric treatment. One would be naïve to use this information to argue either that only lunatics buy television licences or to suggest that purchase of a TV licence is a cause of mental illness. We know quite well that for many years the number of television sets in use showed a steady increase, associated with increasing affluence and changing tastes in entertainment. Over the same period the strains of modern living have led to more mental disorders and this, together with a more freely available medical service and a greater consciousness of the benefits of medical help, has led to a steady increase in the numbers of patients seeking psychiatric treatment.

Is it not possible that some quite separate reason might lead to high smoking and lung cancer going together without any cause-and-effect relationship? For example, may there be certain genetical traits that make a person more susceptible to lung cancer and also of a highly strung disposition and therefore inclined to smoke to soothe their nerves?

Indeed a person of no less eminence than that statistician turned geneticist R. A. Fisher (formulator of the analysis of variance) argued along lines like these. The statistical evidence alone cannot exclude such possibilities.

The role of tobacco as a cause of lung cancer was only established on a reasonably firm basis when certain carcinogens – cancer-producing substances – were isolated in tobacco; it was also shown that they could be absorbed into the system more easily from cigarettes than from cigars or pipe tobacco. Further researches have shown these substances to be more specifically related to tar content of tobacco. While statistical evidence suggested a link between smoking and lung cancer, it was bio-chemical evidence that established that the link was almost certainly there – although some eminent scientists still remain unconvinced.

Correlation and the myth of respectability

Far too many scientific papers quote the value of a statistic called the *correlation coefficient*, which is a measure of association between two variables. Two important points about it are often forgotten. Firstly, it is only a measure of linear or 'straight-line' association. Secondly, a high correlation coefficient, although suggesting a near linear relationship, does not necessarily indicate cause and effect. Relationships not of the straight-line type may give low values of the correlation coefficient and we shall see that low values of the coefficient need to be interpreted with care.

We use more 'algebra' in this chapter than elsewhere in this book because correlation and regression formulae become very complicated in words – and here's hoping algebraic notation will help rather than hinder understanding.

If you do find the going too heavy it is possible to omit the remainder of this chapter at a first reading without hindering your understanding of most of the remaining chapters. (The only later part that is hard to follow without studying this chapter is the section in Chapter 11 that deals with regression on the computer, pages 196–8.)

What is a correlation coefficient? Suppose that we measure two variables, x and y, on two different experimental units, numbers 1 and 2; we can call the measurements on the first unit (x_1, y_1) and those on the second (x_2, y_2). We can carry out this operation on any number of units, say, n, getting a set of n paired observations $(x_1, y_1), (x_2, y_2), \ldots, (x_n, y_n)$. Each unit might be a schoolboy, x representing height and y age. Or each unit might be an industrial worker, x representing his weekly wage and y the amount per week he spent on fuel. The unit might be a calendar year, x representing annual tonnage of freight carried by rail and y the corresponding road tonnage.

If we call the mean of the x values $\bar{x}$ and of the y values $\bar{y}$, we may calculate the deviations of each paired observation from these means; that for the ith unit would be

$$(x_i - \bar{x}, y_i - \bar{y}).$$

These deviations can be represented by n points on a graph; the intersection of the axes represent the position of $\bar{x}, \bar{y}$ relative to the deviations. Doing this for a particular set of data gives a picture of the scatter of our observations.

One type of scatter pattern is illustrated in Figure 8.

Each point in Figure 8 represents a value of $(x_i - \bar{x})$ measured in the horizontal or x-axis direction and a value of $(y_i - \bar{y})$ measured in the vertical or y-axis direction. The x values represent the volume of liquid fertilizer supplied to each of n strawberry plants grown in individual pots and the y values represent the weight of the crop produced by each plant.

The axes divide the figure into four quadrants numbered as indicated. Most of the observations lie in the first and third quadrants, only a few in the second and fourth. The observations in the first and third quadrants indicate values of x above the mean

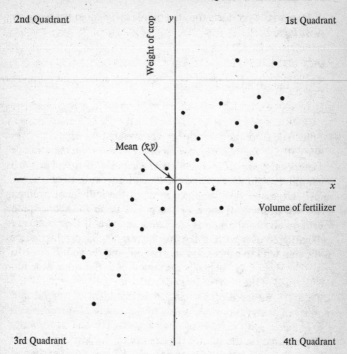

Figure 8 Scatter diagram for positively correlated variables

associated with values of y above the mean, and values of x below the mean associated with values of y below the mean.

It is convenient to introduce a shortened notation for the deviations from means for each unit. We shall write X_i and Y_i for the deviations for unit i, i.e.

$$X_i = x_i - \bar{x}, \quad Y_i = y_i - \bar{y}.$$

For all points in the first quadrant X_i and Y_i are both positive and therefore their product, $X_i Y_i$, is also positive. In the third quadrant X_i and Y_i are both negative so that their product is positive. For the few points in the second and fourth quadrant one deviation is positive and the other is negative so that their product is negative.

If we sum the products for all n points and denote the result by S_{XY} we have

$$S_{XY} = X_1 Y_1 + X_2 Y_2 + \ldots + X_n Y_n.$$

Using the summation sign Σ we may write this more shortly as

$$S_{XY} = \Sigma X_i Y_i.$$

Mathematical notations really are a convenient shorthand!

We call S_{XY} the *sum of products of deviations from the mean* or sometimes the *sum of products about the mean*. If the points fall predominantly in the first and third quadrant (as in Figure 8) the sum is large and positive since most of the individual product terms are positive. If most of the points lie in the second and fourth quadrants the sum is large and negative. If the points are well scattered over each of the four quadrants, the product terms in the sum tend to cancel one another out and the sum is close to zero. The size of S_{XY} depends also upon both the value of n and the 'spread' of the observed x and y values.

To get a measure of association independent of spread and number of observations, we divide by a measure of spread which also increases with number of observations. The one customarily taken is based on the sum of the squares of the deviations of the values of x and y from their means, i.e. ΣX_i^2 and ΣY_i^2; we in fact divide by the square root of the product of these two measures. The resulting quantity, usually written r, is called the *correlation coefficient*.

In words, understanding all deviations to be from the mean, we may write

$$r = \frac{\text{sum of products of deviations of } x \text{ and } y}{\text{square root}\left\{\begin{array}{l}\text{sum of squares of deviations of } x \text{ multiplied}\\ \text{by sum of squares of deviations of } y\end{array}\right\}},$$

or in more compact algebraic symbolism

$$r = \frac{\Sigma X_i Y_i}{\sqrt{\Sigma X_i^2 \Sigma Y_i^2}}.$$

If we write $S_{XX} = \Sigma X_i^2$ and $S_{YY} = \Sigma Y_i^2$ we may write r as

$$r = \frac{S_{XY}}{\sqrt{S_{XX}S_{YY}}}.$$

The three forms of r written above all mean exactly the same but each is a little more compact than its predecessor.

When we calculated sums of squares of deviations from the mean in Chapters 4 and 5 we used rather different computing formulae. Here too we may express r in another form which makes calculation easier. It is

$$r = \frac{\Sigma x_i y_i - \dfrac{(\Sigma x_i)(\Sigma y_i)}{n}}{\sqrt{\left\{\left[\sum x_i^2 - \dfrac{(\Sigma x_i)^2}{n}\right]\left[\sum y_i^2 - \dfrac{(\Sigma y_i)^2}{n}\right]\right\}}}$$

This formidable formula has close analogies with the formula for getting sums of squares of deviations in a t-test or an analysis of variance. The denominator terms are exactly analogous, and the numerator can be expressed in words as

sum of products of x_i and $y_i - \dfrac{\text{product of sum of } x_i \text{ and sum of } y_i}{\text{number of observations}}$.

We have given the formula for r in several different notations (some text books manage even more) but they all are algebraically equivalent and lead to the same numerical results.

Table 26 gives some data for the amounts of inorganic bromine (x) in soil plots of a standard size in micrograms per millilitre and the number of flowers per carnation plant (y), as an average over 30 plants grown on each plot. The data are very similar to some real experimental data reported by Kempton and Maw (1974) but have been simplified to make calculations easier.

The data are also shown graphically on Figure 9. It hardly takes a statistician to see that if we want to grow good carnations we should choose a low-bromide soil. If we were asked to put a straight line on Figure 9 to fit the data as well as possible it would be somewhere near the broken line shown on that figure. Clearly there is a strong but not perfect linear association between bromine level and number of flowers – the higher the bromine level the fewer the flowers.

Table 26 *Toxic effect of inorganic bromine on carnations*

Inorganic bromine in soil (micrograms per millilitre)	Number of flowers per plant (average over 30 plants)
x	*y*
3	3.2
4	2.9
6	3.7
7	2.2
8	1.8
10	2.3
12	1.7
15	0.8
16	0.3

Sum	81	18.9
Sum of squares	899	49.33
Sum of products of corresponding *x* and *y*		133.4

It can be established theoretically that if there is a perfect linear association between *x* and *y*, that is, if all the points in a diagram like Figure 9 lie exactly on a straight line, then *r* takes the value $+1$ or -1. It takes the value $+1$ if the line slopes upwards from left to right; that is if low *x* goes with low *y* and high *x* goes with high *y*. It takes the value -1 if high *x* goes with low *y* or vice versa; then the slope is downwards from left to right. If there is complete scatter, i.e. no association between *x* and *y*, *r* takes a value near zero. We shall see in the next section that certain types of association between *x* and *y* also give values of *r* near zero. The correlation coefficient never has magnitude greater than 1.

Using the data of Table 26 and the computational form of the formula for *r* (page 129) we get

$$r = \frac{133.4 - \dfrac{(81 \times 18.9)}{9}}{\sqrt{\left\{\left(899 - \dfrac{(81)^2}{9}\right)\left(49.33 - \dfrac{(18.9)^2}{9}\right)\right\}}}$$
$$= -0.91.$$

This value indicates a fairly high degree of linear association between *x* and *y*. Given certain assumptions, a test exists to

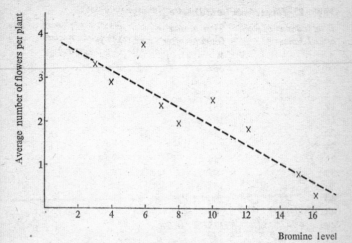

Figure 9 Relationship between soil bromine and flower numbers

decide if a correlation coefficient calculated from a 'sample' of observations indicates a *population* coefficient other than zero. This test assumes normality and it is not particularly robust to departures from this. Tests of significance for correlation coefficients are even less useful than most tests of significance, the limitations of which are evident. When correlation coefficients have any relevance it is nearly always more useful to study linear regressions, as we do later in the chapter.

When it's sunrise in Dundee

This example shows that the correlation coefficient may be near zero even when two variables are closely related. The data in Table 27 give the times of sunrise at Dundee, Scotland, throughout the year. The variable x represents the time of year, measured in weeks from 1 January, and y the time, in minutes after 3 a.m. Greenwich time, at which the sun rises at date x. Successive x values corresponding to 1 January, 4 weeks later, 8 weeks later and so on.

Table 27 *Times of sunrise at Dundee*

Date (number of weeks after 1 January)	Time of sunrise (minutes after 3 a.m. GMT)
x	y
0	347
4	317
8	253
12	181
16	107
20	48
24	21
28	44
32	95
36	152
40	209
44	269
48	325
52	347

Figure 10 is based on these data and illustrates a clear relationship in the form of a smooth curve. But what happens when we calculate correlation coefficients? First let us consider the data for weeks 0–24, roughly the period January to June. The correlation coefficient, calculated by the methods already given, is -0.993. It correctly reflects the fact that as the year progresses from winter to summer the time of sunrise becomes earlier. The high value of r implies that there is a strong measure of linear association between x and y although Figure 10 shows some slight curvature in the relationship over this period.

Using *all* the data in Table 27, i.e. that for the whole year, we find the correlation coefficient $r = 0.064$, a value not far from zero. It would be ludicrous to conclude that knowing the date tells us nothing about the time of sunrise! Yet it is surprising how often scientists regard a small correlation coefficient as an indication of no association at all between variables. In this example taking half the data indicates some linear association. Taking all the data indicates virtually no *linear* association. In truth there is a *curved* relationship. A study of Figure 10 makes it plain why these contrary results are quite logical. Think where $\bar{x}$ and $\bar{y}$ would be!

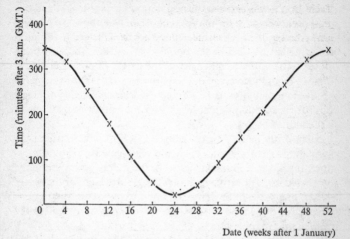

Date (weeks after 1 January)

Figure 10 Sunrise at Dundee

The correlation coefficient between an isolated pair of variables is one of the more useless statistics. There are, however, important uses for groups of these coefficients; we may obtain them if we have a large number of variables and calculate the correlation coefficient for each pair. The applications involve techniques that form part of a branch of statistics called *multivariate analysis*. They require advanced statistical theory and are widely used in the interpretation of experimental results in psychology and many other fields; biological applications, for example, are described in a book with the somewhat forbidding title of *Multivariate Morphometrics* by Blackith and Reyment (1971).

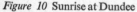

Linear relationships between variables: regression

In 1886 the geneticist Frances Galton noted that tall parents tended to have tall children and short parents tended to have short children, not in itself a very profound observation. Considing eldest sons he also noted a more subtle relationship. He calculated the average height of the two parents, the mid-parent height. He found that the average height of eldest son for a given mid-parent height was *between* the population average height and this mid-parent height. If the mid-parent height was above the population average, their offspring tended to be shorter; if the mid-parent height was below the population average, their offspring tended to be taller. In other words the height of offspring tended to move from the average of the parents towards the average for the whole population. If he plotted the average heights of eldest sons for various mid-parent heights he got a good straight line fit to the plotted points.

Because the heights of offspring tended to shift towards the population mean Galton called the phenomenon *regression* towards the mean, and the line of fit to the points on his graph the *regression line*. The term *regression line* is now used for the *line of best fit*, in a sense to be described later, to any set of data points irrespective of whether or not there is a regression towards the mean in the Galton sense.

There is hardly a statistical problem in the perfect straight line relationships met in physics, e.g. Hooke's law or the voltage–current relationship for a circuit of fixed resistance in accord with Ohm's law. Statistical problems rear their head in approximate relationships such as those implicit in Figure 8. Data resulting in plots of this type may be generated in a number of ways. Galton, for example, could not control the heights of parents or offspring and simply had to take what figures were available. For the data in Table 26 the fact that the given bromine levels are all integers might suggest that the soil in each plot had a known concentration of bromine incorporated in it and that that concentration was pre-specified by the experimenter, the pre-specified values being chosen to cover a range of interest to him; but without further evidence we could not be positive about this, another possibility being that the measuring equipment used was only capable of recording concentrations to the nearest unit. The data alone cannot tell a statistician which of these alternatives corresponds to reality. A competent statistician would ask the experimenter for the correct explanation. Fortunately, in practice, although strictly speaking the best method of fitting a straight line to the data would be slightly different in each case, the answers obtained would not be very different.

We tend to use the same method of fitting even though straight line relationships have different logical statuses. Sometimes they represent underlying law-like relationships (such as Hooke's law), often obscured by errors or noise so observations do not lie exactly on a straight line; such errors may be present in one or both of the variables. Noise may represent inaccuracy in measuring instruments but in many problems, particularly where biological material is concerned, departures from a law-like relationship are more likely to be attributable to a variation superimposed upon the orderly pattern implied by the relationship itself.

For example, if we measure arm length and leg length of adult human males we find that people with long arms generally have long legs and those with short arms have short legs. A plot of arm length against leg length yields points more or less on a straight line. However, points (corresponding to individuals) deviate slightly from a straight line because genetical or environmental factors upset the idealized relationship.

In many situations with a clear-cut cause-and-effect relationship we may be interested in the way the mean of one variable changes as we change the value of the other variable. This was essentially Galton's interest. Another example of this kind is provided by a system of school medical examinations where the ages and heights of children are recorded annually and we are interested in the average heights of children of different ages. Let us suppose that each child is examined on or very near his birthday. For children of a given age, say 11 years, there is a wide spread of recorded heights, but it is fairly well established that within the age range from 6 to 12 years the average heights of children vary linearly with age. The heights of a group of children of the same age represent a sample of all children of that age; by taking samples at several different ages and fitting a straight line as best we can to the height means for each age we obtain an estimate of the population mean height at any other age *within the range of our observations*.

In this example we have made one simplification that does not always accord with reality. We have assumed that the medical examination was carried out on or near to each child's birthday; this implies we know the exact age when the measurement was made. In practice medical examinations are not tied to the child's birthday. We often find in medical records a child's age recorded simply as, say, seven, for any child who has celebrated his seventh birthday but has not yet reached his eighth. Thus any age between exactly seven years and seven years 364 days is recorded as seven. Distortions due to this type of data limitation, known as *grouping by age*, on the fitting of straight lines have not been completely studied even at the theoretical level, but there are practical devices that reduce the effect of such distortions.

In this section we have indicated some of the different ways approximate straight-line relationships might arise; there are interesting theoretical problems in the apparently simple task of fitting a straight line to observations, but we shall by-pass these.

A line of best fit

It helps to think about our model in terms of the age and height relationship we introduced in the last section. We again assume that heights are actually measured on each child's birthday. A typical situation is illustrated in Figure 11.

In Figure 11 individual heights are shown by dots. The crosses represent the mean height of all boys of each given age, i.e. the mean of the dot values corresponding to boys of that age. The line in the figure represents the line of *best* fit to the crosses, in the sense described below.

In theory one can fit a straight line to as few as three points and get a measure of how well that line fits the small amount of data. In practice that is futile and the line obtained may differ considerably from one based on a larger number of observations. The situation is typical of estimating something from a sample; we are trying to estimate the line we would get for 'all' boys in the given age range from observations on just a few.

We put the word 'all' in quotes because we need to give a little thought to the population we are considering. It may well be that the line of best fit for, say, Chinese boys will be different from that for Indian or American or Dutch children.

When we consider the problem of fitting a line in the frame-

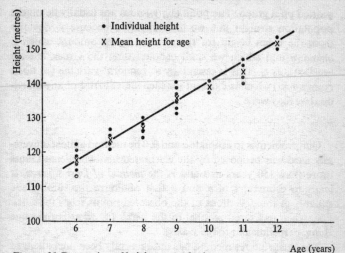

Age (years)

Figure 11 Regression of height on age for boys

work of estimation and hypothesis-testing, two key points are whether a straight-line is a reasonable thing to fit, and if it is, how good an estimate do we get of the ideal *population* straight line.

In mathematical terms we are proposing a mathematical model which accepts the straight-line hypothesis. It says that the mean height, y, for any age, x, in the range six to twelve years is given by an equation of the form

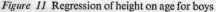

$$y = \alpha + \beta x,$$

where, α (alpha) and β (beta), are 'population' parameters or constants to be estimated from the data or sample values. We have adhered to the convention of using Greek letters for population parameters. The constant β is the *slope* of the line and measures its steepness. If β is positive the line runs upwards from left to right as in Figure 11; if β is negative it runs downwards from left to right as in Figure 9. The greater the numerical value of β the steeper the slope. The constant α tells us where the line cuts the y-axis.

The observation of an individual child's height, y_i, and his age, x_i, are represented algebraically as a point (x_i, y_i) which may be plotted on a graph. The point (x_i, y_i) does not usually lie on the population straight line we have specified because y_i deviates from the mean height for the given x_i by an amount which is *unknown* and which we shall denote, using the Greek letter ε (epsilon), as ε_i. This is essentially a 'random' variable (or noise component) which is different for each individual. For any individual we may write

$$y_i = \alpha + \beta x_i + \varepsilon_i.$$

Our problem is to estimate α and β. The method almost universally used was proposed by the celebrated mathematician Gauss more than 150 years ago and is the *method of least squares*. It leads to estimators of α and β that minimize the sum of the squares of the deviations of the observed points from the fitted line, when measured parallel to the y-axis (or what is the same thing, perpendicular to the x-axis).

The least squares method has until recently been unchallenged and theoretical developments in statistics have given further justification for its use providing certain assumptions about the ε_i are fulfilled. In essence these assumptions are that the ε_i have the same normal distribution and that their mean is zero. Their distribution does not depend on the value of x_i and the value of the ε_i for any one observation does not influence the value for any other observation.

The method of least squares has been generalized to deal with situations where the assumptions do not hold. Recently there has been interest in more robust methods that give very similar results to least squares when appropriate, and sensible results in other cases. Violations of assumptions are more common in industrial situations or the social sciences than in the natural sciences.

It is convenient to write a, b for our estimators of α, β obtained by least squares. They are the values of a and b that minimize

$$U = \Sigma(y_i - a - bx_i)^2$$

where, as usual, Σ means we are to take the sum over all pairs of observed values (x_i, y_i). One may obtain the values of a and b that make U as small as possible by differential calculus. For those familiar with calculus the procedure is straightforward. It can be shown that the expression for b can be put in words as

$$b = \frac{\text{sum of products of deviations of } x_i \text{ and } y_i}{\text{sum of squares of deviations of } x_i}.$$

Using the symbolism introduced when considering correlation coefficients we can write this more concisely as

$$b = \frac{\Sigma X_i Y_i}{\Sigma X_i^2} \quad \text{or} \quad b = \frac{S_{XY}}{S_{XX}}.$$

Again there is a more useful form for computation:

$$b = \frac{\Sigma x_i y_i - \dfrac{(\Sigma x_i)(\Sigma y_i)}{n}}{\Sigma x_i^2 - \dfrac{(\Sigma x_i)^2}{n}}.$$

Comparison of the expression for b with that for r shows that they are related by the formula

$$b = r\sqrt{\frac{S_{YY}}{S_{XX}}}.$$

Unlike r, which is confined to the range -1 to 1, b may take any value. The value of a is given by

$$a = \bar{y} - b\bar{x}.$$

If r has already been calculated all the heavy work has been done for calculating a and b, and between them they are much more informative than r. We illustrate the calculations using the data and the sums of squares and products given in Table 26 for the carnations. Using the 'computing' formula for b given above we see that

$$b = \frac{133.40 - \dfrac{(81 \times 18.9)}{9}}{899 - \dfrac{(81)^2}{9}}$$

$$= -\frac{36.7}{170}$$

$$= -0.216,$$

and since $\bar{x} = \dfrac{81}{9} = 9$ and $\bar{y} = \dfrac{18.9}{9} = 2.1$ it follows that

$$a = 2.1 - (-0.216) \times 9 = 4.044.$$

Thus the least squares line of best fit has the equation

$$y = 4.044 - 0.216x.$$

This is the line in Figure 9. To what use can the straight line be put? One use is for prediction. For example, we might want to know the 'expected' average number of flowers per plant when $x = 11$. It is intuitively reasonable to get a point estimate of this by putting $x = 11$ in the least squares line, i.e.

$$y = 4.044 - 0.216 \times 11 = 1.668.$$

Now, since the original values of y were given to one decimal place only we cannot expect estimates derived from them to be accurate to three decimal places! Therefore we should round off to one decimal place and take 1.7 as our best point estimate of the average number of flowers when $x = 11$. We must remember that this is *just* an estimate. In another experiment we would almost certainly observe different y values even if the bromine levels x were the same. One would hope to get a similar general pattern of results to that depicted by Figure 9, but the estimates a, b of α, β would be different and hence we would get a different estimate of y when $x = 11$.

Since

$$a = \bar{y} - b\bar{x}$$

the estimated line may be written

$$y - \bar{y} = b(x - \bar{x})$$

and the estimated line always passes through the point $(\bar{x}, \bar{y})$. This follows because the equation is obviously 'satisfied' if we put $x = \bar{x}$ and $y = \bar{y}$, as both sides are then zero.

A student of physics will recognize the point $(\bar{x}, \bar{y})$ as the centre of mass of a physical system with unit (or equal) masses at each of the points (x_i, y_i).

If we do another experiment with the same bromine levels (x values) different y values result, owing to biological variations in the plant material and uncontrolled environmental changes. In the new experiment $\bar{x}$ is unaltered but $\bar{y}$, as well as the individual y_i, is different. The estimate b of β also changes from one set of observations to another. Bringing $\bar{y}$ to the other side of our equation it is easily seen that our least squares regression line may be written

$$y = \bar{y} - b(x - \bar{x}).$$

The form of this equation shows that changes in our estimator b of β have no effect on our estimate of y when $x = \bar{x}$, for then we always have $y = \bar{y}$. However, the more x differs from $\bar{x}$ the greater the influence of the particular value of b upon our estimate y corresponding to a given x. In other words when an x is far removed from $\bar{x}$ our estimate y is very sensitive to changes in b. The situation is illustrated in Figure 12; crosses represent the same points as those in Figure 9 and the dots represent a further set of nine observations based upon the same x values but with different average numbers of flowers per plant (y). The least

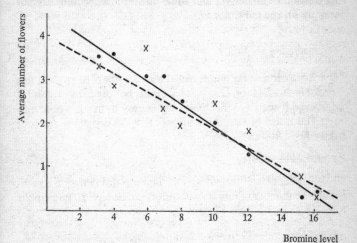

Figure 12 Estimated regression lines for two samples

squares lines of best fit are shown by a broken line for the first set of points and by a continuous line for the new set. Clearly the lines are closer together near their centres than they are at some distance from their respective $(\bar{x}, \bar{y})$, showing that sampling variation has had a greater effect upon extreme estimates than upon those made near the 'centre of mass'.

Not unexpectedly, the greater variability as we move away from $(\bar{x}, \bar{y})$ reflects itself in the expressions for confidence limits and confidence intervals for the 'true' or population expected value of y corresponding to each given x. The confidence limits are more complicated than any we have met so far; they have to take into account not only sampling variation in the observations themselves but also the fact that this exhibits itself again in our estimates of α and β. The reader who wishes to delve more deeply into these matters should consult the chapter on regression that appears in most statistical text books or consult a more specialized book such as Draper and Smith (1966), Sprent (1969) or Daniel and Wood (1971). The first two of these invoke more advanced mathematics than this book.

Confidence limits in regression are extremely important. Regression estimates without some measure of accuracy are not very useful and may mislead.

Overstepping the limits

Our discussion so far has assumed that the linear or straight line model is satisfactory. One point we must be very careful about is that even if such a model looks satisfactory it may not hold outside the range of our observations. For the carnation data our line of best fit was

$$y = 4.044 - 0.216x.$$

What happens if we put $x = 20$ and use this equation to estimate the number of flowers with this bromine level? The estimate is

$$y = 4.044 - 0.216 \times 20,$$
$$= -0.276.$$

The expected average number of flowers is negative! What nonsense. Clearly the linear relationship, if it is valid and useful at all, cannot extend beyond the point at which y is reduced to zero. (Extending a relationship beyond the range of observed values is called *extrapolation* by mathematicians; it is a dangerous game for the statistician to play.) Yet it is hard to see, just by looking at the data, any simple curve other than a straight line giving a reasonable fit to the carnation data.

One assumption we needed to justify the use of least squares was that the variation represented in the ε_i had the same pattern or distribution for all x_i. This means that the scatter of individual points about the line is of the same order of magnitude for all x_i. Now in practice measurements may be made on organisms of very different sizes; in these circumstances the scatter about the line of best fit is very often greater for larger organisms than for smaller ones.

Usually such a dependence of the amount of scatter makes little difference to point estimates of α and β; we can ignore it and use ordinary least squares estimation. It does, however, make confidence limits for estimates rather senseless unless we modify our expressions to take it into account.

Another assumption that frequently breaks down is that all 'departures' from the ideal straight line are in the y-direction. When both x and y may deviate from an idealized value specified by a law-like relationship we have the *errors in variables* model and determining or estimating the true relationship in these circumstances is often called obtaining a *functional* or *structural* relationship. The situation is complicated and the field a happy hunting ground for theoretical statisticians, and one not without practical importance.

Leaving the straight and narrow

Figure 10 represents just one situation where there clearly exists a relationship between variables that is not a straight line. If we are lucky theoretical considerations suggest the most appropriate curve to fit. The principle of least squares may still be applicable, but in general we no longer get simple explicit expressions like those for the estimators a and b of α and β in the straight line situation. You may not have thought the formula for b was simple – but relative to those associated with more complex curves it was.

Sometimes a problem which is not originally one of fitting a straight line can be turned into one. For example if we want to fit a curve of the form

$$y = \gamma x^{\beta}$$

to some observed data – something biologists may want to do – and we don't know γ (Greek gamma) and β we can go about it this way. If we take logarithms the above relationship becomes

$$\log y = \log \gamma + \beta \log x.$$

If we put $\log y = Y$, $\log x = X$ and $\alpha = \log \gamma$ we get

$$Y = \alpha + \beta X$$

and we have a linear relationship.

To justify the use of least squares to estimate α and β we assume that the error is such that, for observed values x_i and y_i for which $X_i = \log x_i$ and $Y_i = \log y_i$, the model is

$$Y_i = \alpha + \beta X_i + \varepsilon_i,$$

where the ε_i satisfy the normality and homogeneity assumptions. In practice these assumptions are often not justified and the problem then has complications, and the advice of a trained statistician is needed.

In the above case we found a straight line relationship between $\log x$ and $\log y$. Other models give rise to straight line relationships either between y and $\log x$ or between $\log y$ and x. It is often useful to plot these variables against one another in a preliminary study of data to see if a straight line relationship may exist in one of these cases.

Not all relationships between pairs of variables can readily be transformed to straight lines. A curve often fitted to data, usually because it seems to provide a reasonable fit rather than upon any strong theoretical grounds, is a polynomial, which is used in *poly-*

nomial regressions. A typical example is the *third degree* polynomial which has the equation

$$y = \alpha + \beta x + \gamma x^2 + \delta x^3$$

where δ (delta) is a further Greek letter denoting a parameter. The term third degree refers to the '3' in x^3, the highest power of x appearing in the polynomial. (It has nothing to do with questionable police methods.) Again we may use the method of least squares to estimate the parameters, but we shall see that the polynomial regression is a special case of multivariate regression (discussed in the next section).

In one situation the fitting of polynomial regressions is especially easy. This is the case where all the x values are equally spaced, e.g. if they are 2, 4, 6, 8, 10, 12 or -5, -2, 1, 4, 7, 10, 13, 16. Details are given in most text books that describe the fitting of polynomials to data.

Multivariate regression

Let's think for a moment about the relationship between weight and age in schoolboys. Older boys generally weigh more than younger boys. There may be a relationship between the two variables that is not far from linear, but deviations from the fitted line are likely to be quite large for some individuals. A boy who is taller than average for his age may weigh more than one of average height; similarly if he is fatter than average he weighs more.

As a first approximation it is not unreasonable to assume that height and fatness (as measured by waist girth for example) might have some sort of additive effect upon weight over and above that of age.

Therefore we hope that a predictive model that also considers height and perhaps waist girth gives a better formula for estimation purposes. So we estimate the parameters α, β, γ, δ in a model of the form

$$y = \alpha + \beta x + \gamma z + \delta u$$

where y represents weight, x represents age, z represents height and u waist girth. The regression problem is then to estimate α, β, γ, δ given, say, n individuals for whom we have made the observations $(x_i, z_i u_i, y_i)$. Again, using appropriate assumptions about departures from our model – normality and homogeneity – our old friend least squares comes to the rescue. With more than a few observations and a few variables a computer is virtually a 'must' to carry out the computations; indeed, a modern computer fits regressions so quickly that many people fit them without considering whether or not they make sense. A model with several variables is usually called a *multivariate* regression model, a generalization of the term *bivariate* which means a case with two variables. Some statisticians prefer the name *multiple regression* rather than multivariate, reserving multivariate for an even more general situation.

The example we have just considered is a *linear* regression model, but it is *not* that of a straight line. In mathematics *linear* has a more general meaning than straight line, its meaning when there are only two variables. With three variables a linear relationship represents a plane or flat surface, a flat sheet of cardboard, for example. With more than three variables a linear relationship specifies a *hyper-plane*. The human mind is unable to picture geometrically what a surface looks like in more than three dimensions, but an idea of what is happening may be built up by recognizing a straight line as a space of one dimension (that of the line) embedded in a two-dimensional space, i.e. any plane in which the line lies. When we draw a line on a sheet of paper the sheet of paper is a plane containing the line.

Similarly a plane is a space of two dimensions that exists in a three-dimensional space; a hyper-plane in four dimensions is essentially a three-dimensional space that exists within one of four dimensions, and so on. Our difficulty arises because we cannot visualize a four-dimensional space as we do a one-dimensional space (line), a two-dimensional space (plane) or the real world three-dimensional space in which we live.

In the last section we mentioned that bivariate polynomial regression can be regarded as a special case of multivariate regression. The equivalence becomes immediately obvious if we put

$z = x^2$ and $u = x^3$ in a four-variate linear regression equation of the type we have been discussing.

Multivariate regression is almost too popular, and a lot of people get into difficulties with it. Often a large number of variables – sometimes 10, 20 or even 50 or more – are measured on a number of individuals or occasions. For example each day at a meteorological station one might measure maximum and minimum air temperatures, ground temperature at several soil depths, air temperatures at specified times, pressure, humidity, rainfall, wind velocity and direction, soil moisture, and so on. In social surveys it is often easy to get a large number of measurements on each unit and there is a tendency to collect data just in case it is useful some day.

Trouble starts when somebody with a lot of measured variables gets hold of a computer program able to fit multivariate linear relationships involving all or nearly all of a vast number of measured variables in a few minutes – or even seconds. Doing this indiscriminately is almost certainly a waste of computer time and can mislead completely. We shall explain why in Chapter 11 where we have more to say about computer programs.

Comparing regression equations

We have talked about using regression equations for prediction and for describing law-like relationships. Sometimes one is also interested in comparing several relationships arrived at in different circumstances. We shall confine our discussion to the bivariate case for simplicity.

Suppose we have established a linear relationship between age and height for schoolboys in London. Does this relationship also hold for boys in Aberdeen? Or Berlin? Or San Francisco? Or Hong Kong?

One way to find out is to take samples of boys from each of these cities and fit regression equations to each set of data. We find our estimates of α, β are different in each case. Is it reasonable to assume that they all estimate the same α and β? Can the differences in our estimates be attributed to sampling variation, or do they imply that α or β or both differ from city to city? This is a typical hypothesis-testing problem.

An old friend comes to our rescue in answering such questions. It is the analysis of variance. For example, tests based upon the analysis of variance can determine whether a set of regression lines estimated from data represent population lines that can be assumed to be parallel to one another or whether they all radiate out from a common point. Further details of such tests are given by Williams (1959) and Sprent (1969).

For your own calculations

We have not in this chapter given any detailed formulae for tests, confidence limits, etc.; the ideas about these developed in earlier chapters carry over to the regression situation although the technicalities are more advanced.

However, we have dealt with point estimates in the bivariate case in detail, so try to get these estimates in the following example.

Table 28 gives the number of rotten oranges, y, in ten randomly selected boxes from a large consignment after they have been kept in storage for a stated number of days, x. Assuming that the regression of y on x is linear (a plot of the data shows this looks reasonable) obtain the equation of the regression line.

Table 28 *Rotting oranges*

Number of days in storage	Number of rotten oranges
x	y
3	6
5	10
8	20
11	22
15	31
18	33
20	39
25	51
27	54
30	63

Use the equation to estimate the expected number of rotten oranges in a box after it has been in storage for (i) 22 days and (ii) 60 days. What reservations, if any, do you have about using these estimates? To save you time in calculating, the sum of the squares of the x values is 3422 and the sum of the products of the corresponding x and y values is 6932.

As usual, if there are any troubles turn to Chapter 13 (pages 228–9) for hints.

8 Stochastic Processes

Queues and other things

In the introductory chapter we looked at queues in a post office as typical of systems where the basic events (the times of arrival of customers and their waits before they were served) were random and unpredictable except within certain broad limits, yet the whole process of operation of the post office had long-term predictability. We called the system a *stochastic process*.

Queues are by no means the only stochastic process. Numbers of animals in a particular region or numbers of fish in a lake also change in a random way with two main contributory elements, changes brought about firstly by births and deaths and secondly by immigration and emigration. In some systems one of these factors may be more important than the other.

Animals of a certain species may hardly ever move from the particular region – perhaps a forest – where they were born, and other animals of the same species may refrain from moving into such 'foreign' territory. Any population changes are then due almost entirely to births and deaths.

On the other hand some fish species spend part of their lives in lakes and part in rivers or the sea so that their seasonal migratory pattern may have great influence on population changes in a region.

To take another example, the fortunes of family surnames have a stochastic element. If a family name is handed down through a stock of prolific breeders the number of people bearing that name increases rapidly. If the same name is borne by a family of shy breeders it is likely to die out.

If we know the probabilities of the bearer of a given name producing 0, 1, 2, 3, 4,... male offspring we can work out the probability that the surname will die out after a fixed number of generations, or the probability that it will die out eventually.

Interesting models represent mathematically the way epidemics spread through a community. These take account of the continuous changes in the numbers of people exposed to infection and the numbers who become immune as the epidemic spreads. Such models can account for periodic outbreaks of complaints such as measles.

Fluctuations on the stock exchange have been modelled as stochastic processes, but nobody has yet produced a model in this way that enables them to make a fortune – or if they have they have kept very quiet about it.

We shall content ourselves firstly with showing some of the less obvious differences between stochastic systems and those systems where things are determined non-randomly. Secondly, we shall look at a simple simulation study – a valuable way of examining the behaviour of a stochastic process.

Waiting for buses that never come by

Suppose we know that a bus passes our front door every 10 minutes but don't remember the exact times when the buses pass. If our city transport system is perfect we know that if a bus has just passed another will be along in *exactly* 10 minutes. Once we have observed the *exact* time one bus passes we can work out precisely when *all* following buses will pass.

Suppose we don't know the times and just go out and wait for a bus. Sometimes we have to wait nearly 10 minutes, on other occasions only a few seconds, before a bus comes. We are equally likely to have to wait any time between 0 and 10 minutes. It is fairly clear that on average our wait is 5 minutes, or half the interval between buses. We call this our *expected* wait (our use of the word *expected* is similar to the use in Chapter 6 (page 118) to imply a theoretical average when we referred to the expected number of people answering a question).

Suppose the buses pass not every 10 minutes but in this pattern: a gap of 15 minutes, followed by one of 5 minutes, another of 15 minutes, another of 5 minutes, and so on, the gaps alternating between 5 and 15 minutes. This is not unusual where buses on two different routes overlap on a common section. The average interval between buses is again 10 minutes, and there are also 6 buses per hour.

What is the expected wait if we go out to catch a bus? At worst we have to wait 15 minutes, at best a few seconds. Our expected wait depends upon whether we come out during a 5-minute or a 15-minute gap. If we come out in a 15-minute gap our expected wait is 7.5 minutes; if we come out in a 5-minute gap it is 2.5 minutes.

At first sight it seems reasonable to take the mean of these two expected waiting times, i.e.

$$\frac{(7.5 + 2.5)}{2} = 5,$$

implying that our overall expected waiting time is the same as before. But wait a minute! The 15-minute gap is three times as long as the 5-minute gap and the chances three times greater that we come out in a 15-minute gap than in a 5-minute gap. We should take this into account in working out the mean waiting time. In more formal terms, there is a probability of $\frac{3}{4}$ that we come out in a 15-minute gap, and a probability of $\frac{1}{4}$ that we come out in a 5-minute gap.

In determining the overall expectation we take a *weighted mean* of the two expectations; that is, the expectations 'conditional' upon our coming out in each gap. The weights which statistical theory tells us to use are the intuitively obvious ones, namely the probabilities of coming out for the bus in the respective intervals. We multiply each conditional expectation (7.5 and 2.5) by the appropriate probability ($\frac{3}{4}$ and $\frac{1}{4}$) and add, giving an expected wait of

$$\frac{3}{4} \times 7.5 + \frac{1}{4} \times 2.5 = 6.25.$$

Thus the new schedule gives an expected waiting time $1\frac{1}{4}$ minutes longer than the 5 minutes of the steady 10-minute schedule.

Even with this deterministic system the expected waiting time depends upon more than the average interval between buses, for this was 10 minutes in each case. Take an even more extreme situation of six buses arriving clumped together at hourly intervals. Six buses per hour again is an average of one bus every 10 minutes, but from the passenger's point of view there is effectively only one bus per hour. If we go out to catch a bus our average waiting time is clearly 30 minutes.

Let's now introduce a stochastic element into the problem. Suppose buses arrive at random in the sense that, for any time interval of fixed length, the probability of a bus arriving is proportional to the length of that interval. It does not depend upon the starting and finishing times of the interval, nor upon when the last bus passed. We are also told that on average a bus passes once every 10 minutes. How long can we expect to wait for a bus?

From our earlier experience of deterministic systems we may suspect the interval will be rather more than 5 minutes when buses pass at intervals of exactly 10 minutes, but rather less than the 30 minutes when all the buses come in a clump. But is it more or less than the 6.25 minutes with alternating 5- and 15-minute gaps?

We can set up a *mathematical model* to work out theoretically the expected waiting time. We can use this model to generate a *realization* of the process, often referred to as a *simulation* as it simulates or mirrors the behaviour of the system over a number of hours or for the arrival of a specified number of buses.

Table 29 is based upon a simulation. It gives the intervals in minutes between the arrivals of 30 successive buses.

The results of a simulation or mock experiment can be looked upon as a sample of observations. Just as the sample mean for real observations does not equal the population mean (a point stressed in Chapters 4 and 5) so the mean interval between buses in the simulation does not equal the process mean of 10 minutes. However, we can hope that a simulation of reasonable length will give an observed mean near the process mean.

The mean interval for the simulation is obtained from Table 29 by adding all the interval lengths and dividing by the total number of buses, 30. The result is 9.73, in fairly close agreement with the process mean of 10.

To obtain the mean waiting time during the period covered by the simulation we note first that when we go out in an interval between buses which is, say, t minutes long, then our expected waiting time is $\frac{1}{2}t$ minutes. To get the mean waiting time over the whole simulation we must generalize the idea of a weighted mean used for alternating 5- and 15-minute intervals; each expected time of $\frac{1}{2}t$ is given a weight, which is essentially the probability that we will go for a bus in that interval. Let us denote the length of the interval between the time we start the simulation and the arrival of the first bus by t_1, that between the arrival of the first and second bus by t_2, that between the second and third by t_3, and in general the interval between the arrival of the $(i-1)$th and the ith bus by t_i.

Table 29 *Intervals between random bus arrivals (Process mean interval 10)*

4	7	1	2	3	20	13	6	15	18
17	0	4	6	6	40	21	7	9	21
4	5	7	0	18	4	4	8	10	12

The probability of a man going out in the interval of length t_i, if he is equally likely to go out at any time during the simulation, is proportional to the length of that interval and may be written

$$p_i = \frac{t_i}{T}$$

where T is the total time of the simulation, i.e. $T = \Sigma t_i$.

To get the overall expected waiting time we take the expected waiting time for each interval weighted by its corresponding p_i. Since the expected waiting time for the interval of length t_i is $\frac{1}{2}t_i$, the overall expected waiting time, E, is

$$E = \frac{\Sigma p_i t_i}{2}$$

where Σ represents summation over all 30 intervals. Since $T = \Sigma t_i$,

whence $p_i = \dfrac{t_i}{\Sigma t_i}$, we get

$$E = \frac{\Sigma t_i^2}{2\Sigma t_i}.$$

We may work this out for the data in Table 29 where the times given are the relevant t_i. Their sum is 292 and the sum of their squares is 4976. Thus

$$E = \frac{4976}{2 \times 292} = 8.52.$$

Thus the expected waiting time is 8.52 minutes, nearly as long as the average interval of 9.73 minutes between buses. Another simulation would give a slightly different mean interval between buses, but usually one not far removed from 10 minutes, and also a different expected waiting time. If we carry out a large number of simulations, or what is much the same thing, carry out a simulation covering a much longer period or a much greater number of buses, we find that the mean interval between buses and the expected waiting time, E, *both* approach 10 minutes.

Most people find this result surprising, but it is not too difficult to establish the result theoretically by calculus. It provides at least a partial explanation of the paradox that a bus company can claim to operate a 10-minute service while passengers complain that they are lucky to wait less than 10 minutes and often have to wait longer. Although a bus company may try to operate a regular 10-minute service, traffic congestion gives a six buses per hour service many of the characteristics of a random process: the average wait for passengers tends to build up from the 5 minutes characteristic of a regular service to something nearer the 10 minutes of the random one. Human nature being what it is, people remember a quarter of an hour waiting in the rain better than the sunny day when the bus came along in two minutes. They soon convince themselves the service is worse than is claimed by the bus company, who in their turn do not always appreciate the influence of 'traffic conditions beyond their control' on how long their customers wait.

This type of stochastic process – where there is a fixed probability of an event occurring in an interval of given length, irrespective of when it starts or finishes – is a *Poisson process*. It arises in a number of practical situations. For example, the distribution of the number of faults per unit length in a manufactured fibre is often random in this sense. Theoretical results enable us to work out the expected number of faults in, say, a thousand yards of fibre, or the probability of more than, say, two faults in a given length. Poisson results are important in providing guarantees and in quality control problems of the sort discussed in the next chapter.

An interesting property of a Poisson process is that past history provides no guide to the future. If buses arrive according to a Poisson process at an average interval of 10 minutes, then the knowledge that three, four or even five buses have gone past in the last 5 minutes tells us nothing about how long we must wait for the next bus. It may follow almost immediately or there may be a long gap. All we do know is that at whatever time we go out our average wait is 10 minutes. This is in sharp distinction to the completely deterministic system of one bus every 10 minutes we started with; for then a knowledge of the time that one bus passes tells us when *all* later buses will pass.

What the doctor ordered

Let's now look at a simple model of another queueing system, the appointment system at a hospital outpatients' department. There are obvious stochastic elements – consultations take different times, some patients arrive late for appointments and a few do not turn up at all. For illustrative purposes we look at a very simplified system where the only stochastic element is the variation in consultation time.

Suppose the hospital makes appointments for patients to see one consultant at 10-minute intervals from 9.00 to 11.50 a.m. inclusive. Suppose that consultations take either exactly 5, 9 or 15 minutes depending upon the type of examination or treatment he has to give.

This pattern of consultation time may not be unreasonable in some types of clinic. Newcomers to the clinic may need either a brief examination (5 minutes) to decide that no further treatment is needed, while newcomers who need treatment require a long consultation (15 minutes). Regular attenders at the clinic for routine treatment may require a 9-minute consultation. From past experience it is known that in the long run one fifth of all patients require a 5-minute consultation, one half require a 9-minute consultation and the remaining three tenths require 15 minutes. The proportions above represent the probabilities that consultations will be of those lengths. The fact that we know these, but not the categories into which individual patients fall, introduces the stochastic element.

Following the method used in the bus example, we multiply each consultation time by its probability and add to get the average or expected consultation time, i.e.

$$E = \tfrac{1}{5} \times 5 + \tfrac{1}{2} \times 9 + \tfrac{3}{10} \times 15$$
$$= 10.$$

Thus the mean time of 10 minutes per consultation is the same as the systematic interval between appointments.

If the doctor is free he will see a patient as soon as he arrives. We shall make the further simplifying assumption that all patients turn up precisely on time – neither early nor late. If a previous consultation is still in progress the patient joins any other waiting patients who are seen in their order of arrival as soon as the doctor is free up to 12 noon. The doctor completes any consultation he starts before noon, but patients still waiting then have to make a fresh appointment.

Our interest is in the average time a patient has to wait to see the doctor; the total time the doctor is idle between consultations awaiting the next patient; and the number of patients forced to make fresh appointments because they have not been seen by noon. In a good system all these should be small.

We must emphasize that we have greatly simplified reality for illustrative purposes by ignoring late arrivals and missed appointments and by allowing only three discrete consultation times rather than a continuous range of times. Our model could be extended to remove these restrictions if we had the appropriate information, e.g. the probability a patient does not turn up, arrives five minutes late, etc. These features complicate the system but do not introduce new basic principles.

For models as simple as this, there is a body of mathematical theory that enables us to work out the averages in the long run and also how these figures vary from day to day. For more complicated systems the theory is less amenable and a useful tool for examining such systems is computer simulation.

In a matter of seconds a computer can simulate the behaviour of 100 or even 1000 such clinics and print out for each the averages we want.

It is essential always to remember that results from these simulated or 'mock' clinics only reflect what happens in a real clinic if the correct assumptions are made in setting up the simulation model.

The real value of simulation becomes apparent in more complicated systems, especially if there is little mathematical theory to aid us. Simulation is also valuable in studying the effect upon the stability of a system of small changes in, say, the interval between appointments or the times taken for consultations.

Even without a computer it is possible to do a few 'pencil and paper' simulations of the appointment system we have described. Indeed, when one is programming a computer to do simulations it can be worthwhile first to do a few by hand as an aid to sorting out the logic needed for the computer program.

We illustrate two simulations of the system we have described in Tables 30 and 31. To determine the lengths of consultations for each patient we have made use of a table of random digits. This consists of a sequence of the digits 0, 1, 2, 3, 4, 5, 6, 7, 8, 9 with obvious random properties: it has no pattern of any sort that enables us to predict the next digit in the sequence, and in the long run each digit appears equally frequently. Each digit has a probability of $\frac{1}{10}$ of being the next digit to appear in the sequence.

Random digits are 'generated' using *random number generators* which vary greatly in sophistication, from drawing numbered marbles from an urn to complex electronic devices often called pseudo-random number generators.

Most of us carry a simple random number generator in our pockets in the form of a humble coin; the results of a sequence of four tosses of a coin can provide a table of random digits.

People who have to make use frequently of random digits usually resort to published tables if only a few are required or else use a computer to obtain random or at least pseudo-random digits. Most statistical tables and some text books contain sets of random digits.

Table 30 *First simulation of a simple hospital appointment system*

1 Random digit	2 Arrival time	3 Consultation length (minutes)	4 Start time	5 End time	6 Doctor idle (minutes)	7 Patient wait (minutes)
5	9.00	9	9.00	9.09	1	0
2	9.10	9	9.10	9.19	1	0
4	9.20	9	9.20	9.29	1	0
9	9.30	15	9.30	9.45	0	0
4	9.40	9	9.45	9.54	0	5
0	9.50	5	9.54	9.59	1	4
1	10.00	5	10.00	10.05	5	0
6	10.10	9	10.10	10.19	1	0
7	10.20	15	10.20	10.35	0	0
2	10.30	9	10.35	10.44	0	5
4	10.40	9	10.44	10.53	0	4
0	10.50	5	10.53	10.58	2	3
7	11.00	15	11.00	11.15	0	0
3	11.10	9	11.15	11.24	0	5
0	11.20	5	11.24	11.29	1	4
5	11.30	9	11.30	11.39	1	0
5	11.40	9	11.40	11.49	1	0
0	11.50	5	11.50	11.55	–	0
				Total	15	30
				Mean wait		1.67

In the simulations in Tables 30 and 31 we use random number sequences from such tables to aid us in assigning consultation times with correct probabilities. It works in this way. Since the probability of any digit occurring at a particular place in a random number sequence is $\frac{1}{10}$, and only one digit may occur at a given place, it follows that the probability of the occurrence of exactly one of a pair of specified digits, 0 and 1 for example, is, by the addition rule for mutually exclusive events (page 31) the sum of the probabilities of each occurring, i.e.

$$\tfrac{1}{10} + \tfrac{1}{10} = \tfrac{1}{5}.$$

Table 31 *Second simulation of a hospital appointment system*

Columns

1 Random digit	2 Arrival time	3 Consultation length (minutes)	4 Start time	5 End time	6 Doctor idle (minutes)	7 Patient wait (minutes)
5	9.00	9	9.00	9.09	1	0
8	9.10	15	9.10	9.25	0	0
7	9.20	15	9.25	9.40	0	5
7	9.30	15	9.40	9.55	0	10
8	9.40	15	9.55	10.10	0	15
3	9.50	9	10.10	10.19	0	20
3	10.00	9	10.19	10.28	0	19
6	10.10	9	10.28	10.37	0	18
7	10.20	15	10.37	10.52	0	17
6	10.30	9	10.52	11.01	0	22
2	10.40	9	11.01	11.10	0	21
2	10.50	9	11.10	11.19	0	20
0	11.00	5	11.19	11.24	0	19
0	11.10	5	11.24	11.29	0	14
8	11.20	15	11.29	11.44	0	9
9	11.30	15	11.44	11.59	0	14
6	11.40	9	11.59	12.08	0	19
–	11.50	–	–	–	–	–

Total   1   242

Mean wait   14.2

One patient not seen

Similarly the probability that exactly one of four specified digits occurs is

$$\tfrac{1}{10} + \tfrac{1}{10} + \tfrac{1}{10} + \tfrac{1}{10} = \tfrac{2}{5}$$

and so on, using the obvious extension of the addition rule to more than two events.

Suppose that we associate random digits 0 or 1 with a 5-minute consultation time; this assigns to it the correct probability of $\frac{1}{5}$. We associate random digits 2, 3, 4, 5 or 6 with a consultation time of 9 minutes, assigning it the correct probability of $\frac{1}{2}$. Finally we associate random digits 7, 8 or 9 with a consultation time of 15 minutes, assigning it the correct probability of $\frac{3}{10}$.

In both Tables 30 and 31 the first column is the random number 'determined' for that patient by taking digits, in sequence, from a random number table starting at an arbitrary point in that table. (It would not do to start always at the beginning of the table and work through systematically because this would always use the same number sequence and all simulations would give the same result.)

Column 2 in each table gives the patients' arrival times; the consultation time, based upon the random number in column 1 and the code indicated above, is in column 3. Columns 1, 2 and 3 may be completed for all patients before entries are made in the later columns. Column 4 gives the time a consultation starts for each patient and column 5 the time it ends. If it ends before the next patient arrives the doctor has to wait and his idle time is given in column 6. A zero indicates that the next patient is seen immediately. Column 7 gives the time, if any, a patient has to wait before his consultation commences. Columns 4 to 7 are most easily filled in one row at a time.

In Table 30, for example, the first patient commences his consultation at 9 and ends it at 9.09. As the next patient does not arrive until 9.10 the doctor has one minute idle time and this is entered in Column 6. As the patient did not have to wait, zero is entered in column 7. In a similar way we can fill in the entries in these columns for the patients arriving at 9.10, 9.20, 9.30 and so on.

At the bottoms of columns 6 and 7 we give the total times the doctor is idle and the mean waiting time for all patients actually seen. We note that for the simulation in Table 31 one patient has to make a fresh appointment. This would become evident at 11.55 when the last but one patient has still not been seen, for even if his appointment were for the minimum time it would last at least until noon.

In the first simulation the system works pretty well. The doctor is only idle for 15 minutes, and the odd minutes off might be well-earned breathers. Most patients are seen promptly with an average wait of under 2 minutes and a maximum wait of 5 minutes.

The second simulation shows a less happy state of affairs. The doctor is hard at it all morning, all except the first two patients have appreciable waits with an average of nearly 15 minutes, and one patient has the inconvenience of a trip to the hospital for nothing. For this last patient we record no waiting time because it will be clear by 11.55 that this patient cannot possibly be seen and he would make a fresh appointment. If a waiting time were recorded, one of 5 minutes would be appropriate.

The two simulations give one pattern that is satisfactory and one that is not. If our model of the system is correct both patterns *could* occur. What is interesting is which is more typical. If something like the situation in Table 31 is rare and the pattern in Table 30 usual, we can regard the system as satisfactory. If the Table 31 situation were commonplace we should be less happy. What has happened in Table 31 is that a run of 15-minute consultations early in the day has put things behind and it has not been possible to catch up.

Is such a run likely? It's not difficult to work out the probability that four consecutive consultations will all be of 15 minutes' duration. The nature of random digits ensures that the length of any consultation is independent of the length of any other consultation. Now the probability of any one consultation being 15 minutes is $\frac{3}{10}$; thus by the multiplication rule the probability of all four consultations being of this length is

$$\tfrac{3}{10} \times \tfrac{3}{10} \times \tfrac{3}{10} \times \tfrac{3}{10} = 0.0081.$$

This is a small probability – less than 1 in 120 – but runs of consultations like 9, 15, 15, 15 or 15, 15, 9, 15 are also awkward. Such sequences all have fairly low probabilities, so that we may expect situations like that in Table 31 to occur from time to time but not often.

A much better idea of the likely general pattern emerges if a large number of simulations are done on the computer. If we do this for our model we find that the situation in Table 30 is perhaps a little better than average, whilst that in Table 31 is very far from typical. The model we have adopted is simple enough for these findings to be checked by a theoretical analysis.

How could we improve the system if a situation like that in Table 31 occurred frequently? Unless extra medical staff could be employed or the time of each consultation reduced, the only practical step seems to be increasing the interval between appointment times from 10 to, say, 12 minutes. This would decrease patients' waiting times and increase the doctor's idle time. We could perform simulations to see whether both of these fell into socially acceptable limits. A side effect of this change would be a decrease in the number of patients for whom appointments could be made, and this would probably result in a backlog in the appointment system.

Self simulation

As an exercise you may like to carry out a simulation like those in Tables 30 and 31 for a slightly different case. Suppose this time there are four possible consultation times, 5, 6, 12 and 20 minutes, with probabilities $\frac{1}{3}$, $\frac{1}{3}$, $\frac{2}{3}$ and $\frac{1}{3}$ respectively. What is the mean or expected consultation time? Simulate the system with the same conditions, i.e. appointments at 10-minute intervals from 9.00 to 11.50.

Table 32 is a short table of random digits. Use these in your simulations starting from different points and using non-overlapping sections of the table each time.

Table 32 *Random digits*

6	3	5	3	4	8	8	2	3	1	5	9	9	1
8	2	6	3	0	6	7	9	6	0	2	9	5	4
7	7	1	4	3	2	0	8	5	8	3	9	0	7
8	7	1	2	5	6	6	5	9	1	3	7	6	1
4	3	5	7	4	5	4	9	6	7	5	8	4	5

Do you think the appointment system is unsatisfactory? If so, how do you think it should be altered?

Suppose the system is modified slightly so that there is a probability of 0.1 that a patient given an appointment will not turn up. How might you modify your simulation study to cope with this?

If you are stuck with this hospital simulation first aid is available in Chapter 13, pages 229–30.

9 Statistics on the Shop Floor

When the machines go wrong

It's a far cry from the day when cheap labour was used to weigh sugar into bags or individually wrap bars of chocolate. Weighing and wrapping of products is now almost entirely mechanized.

Similarly the manufacture of everyday chemicals like paints, detergents, cosmetics, tooth-paste or furniture polish is largely an automated process that depends upon machines grading and mixing various constituents in the correct proportions; but machines do go wrong!

Suppose a sugar supplier has an automatic device to weigh 1 kg of sugar into each bag. If a pin in the machine breaks so that it starts delivering erratically he wants to know this as soon as possible. Before the pin actually breaks there may be signs that something is wrong in the form of delivered weights moving progressively off target or becoming more variable than is usual.

How can statistics come to our aid by providing early warning of such mishaps? The statistical detection methods are very largely based on *quality control* checks. When a continuously operating system is monitored we want a quick warning if something appears to be going wrong. If possible we would like this without the need for elaborate calculations. *Quality control charts* provide a solution in the form of simple graphical methods that may be used by a foreman or technician without any special statistical training to read off warning signs.

Let's illustrate the principle by the simplest of such charts for the machine filling sugar bags. In practice the machine does not deliver the same amount to every bag. Under British law the supplier is required to state the minimum net contents on a package and most other countries have a comparable requirement.

Because of variations in the amount delivered the supplier will set his machine to deliver on average slightly more than 1 kg if 1 kg is his guaranteed minimum content.

How much more depends upon the accuracy of his machine. Suppose that experiments show that it delivers amounts to within plus or minus 1.5 g from the mean value almost every time. It would then be sensible to set the machine at 1001.5 g. This would mean that some lucky customers get 1003 g in a bag whereas some would get only 1000 g, but the manufacturer would stay within the law without giving away too much free sugar.

If he were worried about the amount he gave away he might try to adjust his machine to weigh more accurately or perhaps buy a more expensive machine that gave a smaller variation, say only 0.5 g on either side of the mean delivery. This would be worth his while if the saving in sugar 'given away' to ensure the legal minimum is met more than offsets the cost of the machine. This is a matter of accountancy and economics rather than statistics, and we shall assume that the firm continues to use a machine that gives a variation of 1.5 g on either side of the mean.

We look now at how control charts may be used to see that all is well and in particular that the mean is not slipping away from its setting of 1001.5 g per delivery.

To set up the chart initially a large number of bags are weighed; we check empirically from these data that the weights have a 'distribution' with a mean of 1001.5 g and that it is reasonable to assume that the distribution of weights accords with our old friend the *normal distribution*. We also require some measure of the spread of the distribution. This is provided by another old friend, the quantity s^2 we calculated in Chapter 4 (pages 77–8). Remember, it is the sum of squares of deviations of each observation from the mean, divided by one less than the number of observations. To draw up a satisfactory control chart one would want at this initial stage some hundreds, or if possible even a thousand or more, observations to check normality and get a value of s^2 for future use.

When using the charts we don't need large numbers of weighings to see what is happening. Instead we take a small sample of, say, four bags at regular intervals and calculate only the total or mean weight of these four bags. The total requires just a little less calculation but it doesn't matter which we use providing we have the correct chart.

Suppose s^2 calculated from our large sample turns out to be 0.25. Then s, the square root of s^2 has the value 0.5. Now it can be established theoretically that when all is well or – to use the shop floor jargon – when the process is *in control*, the total weight of four bags should lie within 2 g above or below 4006 g (i.e. 1001.5 × 4) in about 95 per cent of all samples and within 3 g of 4006 g in about 999 in 1000 samples.

The more general formulae applicable to any case where the variability is normally distributed are quite simple. If $\bar{x}$ is the mean weight per bag calculated from our initial large sample and s^2 is also calculated from that sample as the measure of spread, then if we take regular small samples of size n, for about 95 per cent of these the total weight should be between

$$n\bar{x} - 2s\sqrt{n}$$

and

$$n\bar{x} + 2s\sqrt{n}$$

and for about 999 samples in every thousand the total weight should be between

$$n\bar{x} - 3s\sqrt{n}$$

and

$$n\bar{x} + 3s\sqrt{n}.$$

If one wishes to be a little more precise about the 95 per cent and the 999 in 1000, the values 2 and 3 in the above expressions should be replaced by 1.96 and 3.09 respectively, but this complicates the arithmetic with little practical effect. The numerical values we gave for the sugar bag example were obtained by putting $n = 4$, $\bar{x} = 1001.5$ and $s = 0.5$ in these expressions.

For our problem we set up a control chart looking like Figure 13. Basically this chart has on it five horizontal lines. The vertical scale represents total weights for samples of four bags.

The horizontal line at 4006 on the vertical scale is the *central line*. The lines at 4008 and 4004 are the *upper* and *lower inner*

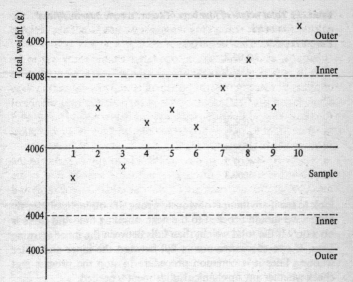

Figure 13 A control chart for sugar bag filling machines

warning lines. We expect in the long run that 95 per cent of the total weights for samples of four will lie between these limits. The lines at 4009 and 4003 are called the *upper* and *lower outer warning lines.*

Suppose that our samples of four are taken at 15-minute intervals. An operator finds their total weight; say it is 4005.2 for the first sample. He plots this on the control chart below the point labelled 1 on the horizontal scale on the centre line. The total weight of the second sample taken 15 minutes later, 4007.1 g, is plotted above the point labelled 2. Table 33 gives readings for a set of ten samples and these have been plotted on Figure 13.

How do we use the chart when we have plotted each point on it? If all points fall between the inner warning lines the rule is clearly leave well alone. There is nothing to indicate anything is wrong. If a point falls between the inner and outer warning lines we need not be too alarmed; we expect this to happen once in twenty times in the long run, but it may be worth having a quick

Table 33 *Total weight of four bags of sugar: sample interval fifteen minutes*

Sample number	Total weight (g)
1	4005.2
2	4007.1
3	4005.6
4	4006.7
5	4007.0
6	4006.5
7	4007.7
8	4008.4
9	4007.2
10	4009.3

look to see if anything is obviously wrong. If nothing is obviously wrong we should proceed to the next sampling time and there is no worry if the total weight then falls between the inner warning lines. If two successive points fall between the inner and outer warning lines it is common procedure to stop the process and check whether any mechanical adjustment is needed.

Should any *one* point fall outside the outer warning lines this is an indication that something may well be wrong – unless a less than one in a thousand chance has come off (compare this with the hypothesis-testing situation!) and the shop floor rule is to stop the process and check for mechanical faults.

Looking for different kinds of faults

The type of chart we have just been describing detects *slippage*, a shift or slip in the average amount delivered by the machine. Control charts can be used for other purposes, such as the detection of increased variability, a fault often associated with a part becoming loose or worn in mechanical devices. Variability may increase without any change in the mean. The sugar bag filler, for example, may continue to deliver 1001.5 g per bag on average, but the variation may grow from between plus and minus 1.5 g to 3 g and then an appreciable number of bags are under the guaranteed minimum. Appropriate charts can detect changes in variability.

The same sample can be used for this detective work as was used for spotting changes in the mean, but we need to weigh each of the four bags in the sample separately.

Making more use of our data – CUSUM *charts*

There's one great weakness about the control charts described so far. The decision whether or not to stop the process at any stage is made on the basis of the current sample point or at most of the last two if both fall between the inner and outer warning lines.

Now clearly we don't want to stop for false alarms, but our small samples may not detect an economically important although not dramatic shift in the mean for quite a while. Can we not in some way accumulate any evidence there might be in successive samples? A technique largely developed since the Second World War helps us to do this.

We'll take the slippage problem with filling sugar bags again. If all is going well we regard our machine as being set to hit a target of 1001.5 g per bag. If the machine goes out of adjustment to the extent that on average it delivers 1002.5 g per bag it behaves as though that were its target. With the original target of 1001.5, points on the control chart like that in Figure 13 tend to scatter above and below the centre line with no obvious trend away from it. If however, the mean shifts upwards there tend to be more points above the centre line than below. Figure 13 suggests that after the first four or five samples were taken the plotted points tend to drift upwards.

We must however not read too much into this, for we saw in Chapter 8, when simulating a stochastic process, that *runs* of events do occur, particularly in the short term. This was illustrated in Table 31 (page 160) where four appointments in succession were all of 15 minutes, an occurrence having probability 0.0081.

When filling the sugar bags is in control it is reasonable to assume an equal probability of a sample point being above or below the target. Thus the probability of six successive points being above the line is $(\frac{1}{2})^6$, or $\frac{1}{64}$.

If there is no slippage successive differences from the target value, added together, tend to cancel one another out – their cumulative sum hovers around zero. This is because they are random variables and have a zero mean value. If there is slippage the differences are still random, but their mean is the difference between the true target value and the actual amount of slippage.

Consider the total of successive deviations from the target value added together. As soon as slippage has occurred, the sum of the deviations from the target value tends to move away from zero. If the amount of slippage is m and we add the deviations for r samples after slippage takes place, the sum of the deviations approachs the value mr.

If we plot the sum of the deviations from target values for successive samples this sum hovers round zero while we are on target, but climbs above or drifts below zero if there is slippage. Sums of deviations are called *cumulative sums*, usually abbreviated to CUSUM.

Table 34 gives the cumulative sums of deviations from the target value of 4006 for the total weights of four bags of sugar. Each entry after the first in column 3 of the table is obtained by adding

Table 34 CUSUM *for sugar samples with target 4006*

Observed weight	Deviation from 4006	CUSUM
4005.2	−0.8	−0.8
4007.1	1.1	0.3
4005.6	−0.4	−0.1
4006.7	0.7	0.6
4007.0	1.0	1.6
4006.5	0.5	2.1
4007.7	1.7	3.8
4008.4	2.4	6.2
4007.2	1.2	7.4
4009.3	3.3	10.7

to the entry above it the deviation in column 2 in the same row as the CUSUM being formed, e.g. the entry 0.6 in row 4 derives from −0.1 + 0.7.

We may plot CUSUM on a chart like Figure 14, where the horizontal or zero line represents the base about which we expect the

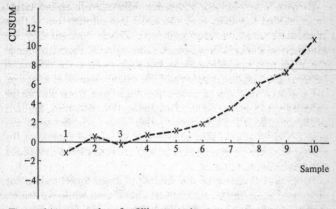

Figure 14 CUSUM chart for filling sugar bags

CUSUM to fluctuate if we are on target, i.e. if the process is under control. We look for trends away from this line as indicators of the process being out of control.

Figure 14 clearly indicates that after the first few points there is a move away from the base line. The dotted line joining successive points is a visual aid to outline the general trend; the line is often put on charts, but it is not essential.

A foreman or controller looking at the chart would become suspicious after observation 7 or 8 that something was going wrong and stop the process, rather than wait for observation 10 as he would using the earlier type of control chart with warning lines.

To remove the subjective element as far as possible we need decision rules analogous to those used with the earlier charts. Note that what is important is the slope of the trend. One step after slippage of size m occurs, the deviation has average value m; two steps after, the CUSUM average value is $2m$; three steps after, $3m$, and so on. Mathematically we say the line has slope m. In trigonometric terms the slope is defined as the tangent of the angle the trend line makes with the base line just like the slope of a regression line. In essence it is the average additional distance the new trend line moves above the base line for each additional sample.

The slope is an important indicator both of the presence and of the amount of slippage. A moment's reflection will show that altering the scales on the two axes in Figure 14 makes the pattern either more clear or less clear. If we squeeze up the vertical scale the trend line is squashed down towards the base line and visual detection of slippage becomes harder. If we expand the vertical scale as we have done in Figure 15, quite small deviations give the appearance of a trend and large deviations may well push the points off the top or bottom of the chart. For sensible visual interpretation some care is needed in the choice of scales; the choice also influences the shape of the V-mask described in the next section.

The choice of vertical scales for best detective work on slippage depends on the variability of observations. If s denotes the estimated population standard deviation obtained in the way we described for ordinary control charts (page 166), a good rule for CUSUM charts for the sum of four observations in our example is

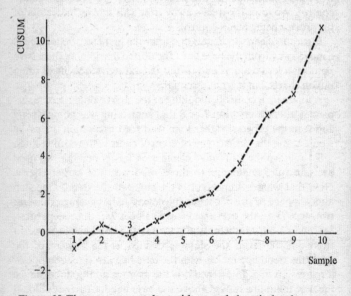

Figure 15 The same CUSUM chart with expanded vertical scale

that 1 unit on the horizontal scale should be approximately equal to $4s$ units on the vertical scale. More generally, if each point on our chart is the sum of n observations one unit on the horizontal scale should be approximately equal to $2s\sqrt{n}$ units on the vertical scale. If the mean, rather than the sum, of n observations is plotted we replace $2s\sqrt{n}$ by $\dfrac{2s}{\sqrt{n}}$.

For a further discussion of CUSUM charts see Wetherill (1969) and Davies and Goldsmith (1972) who go into some technical detail.

The V-mask and decision making

The references just given deal among other things with the way we arrive at decisions using CUSUM charts. There are two commonly used methods. Barnard (1969) suggested a *V-mask* and we illustrate its use in Figure 16. The mask consists of a piece of paper, card or perspex with a V cut out of it. It is placed on the CUSUM chart in the manner illustrated in Figure 16 with the vertex pointing forward and at a designated horizontal distance d in front of the last observed point.

The angles between the base line direction or horizontal and a mask boundary direction (arms of the V) are denoted by θ. The

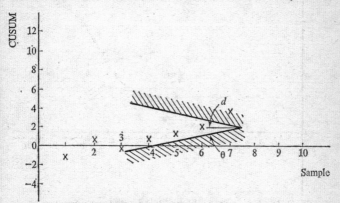

Figure 16 Using a V-mask after the sixth sample

choice of θ depends upon the scales used and the 'behaviour' of the mask depends both upon the values of *d* and θ. Whetherill (1969) describes the way *d* and θ should be chosen. The object is to use the mask in such a way that the process is only stopped if there is strong evidence that it is getting out of control. The rules are that we stop if either of the mask boundary lines intersects the trend path indicated by the sample points.

While we would not stop the process in the situation illustrated in Figure 16, a contrasting situation is given in Figure 17 where we have taken two further samples. Here we would stop because the trend path intersects the *lower* arm of the mask suggesting that the process has moved *above* the target mean. Both Figures 16 and 17 use the data of Table 34.

If one prefers tables to charts another method of using CUSUM invokes a *decision interval scheme*. Whetherill (1969) again gives details. Two values called a *reference value* and a *decision value* or boundary are selected. Sample values are observed until one exceeds the reference value. As soon as the reference value is exceeded CUSUM are formed for (*x–k*) where *x* is the sample value and *k* the reference value. If the CUSUM change sign (i.e. pass through zero) no further CUSUM are formed, but if they exceed the decision value (which is always greater than *k* if the

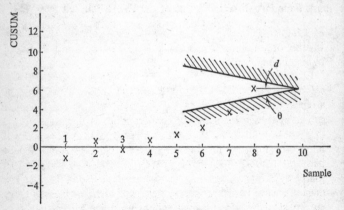

Figure 17 Using a V-mask after the eighth sample

latter is positive and is always negative and greater in magnitude than k if the latter is negative) the process is stopped.

The economics of control charts

The frequency with which samples are taken, the size of sample and whether to use ordinary control charts or CUSUM charts involves economic considerations.

If sampling is destructive there are clear advantages in keeping sample size small and in taking samples no more frequently than is absolutely necessary. How often samples are necessary depends upon how important it is to detect any breakdown quickly. Obviously it is important if a process going out of control leads to production of worthless items.

Whether ordinary control charts or CUSUM charts are most useful depends to some extent on the type of slippage we expect. CUSUM charts are very good for detecting small slippages but not for picking up a sudden fairly large slippage. The slope properties of the trend in CUSUM charts also gives a handy means of estimating the magnitude of a slippage after a few samples.

Sampling inspection plans

In Chapter 1 we briefly referred to the problem of deciding whether or not crates of grapefruit were of an acceptable standard by examining periodic samples.

Let's explore this problem a little further. Suppose the buyer wants to accept a whole consignment only if less than 2 per cent are below standard. Suppose too that he is willing to assume that sub-standard fruit simply occurs at random in the crates. (This may not be realistic for defects due to, say, a fungal disease, for if one fruit is infected the disease may spread leading to pockets of bad fruit. Neither the methods nor the theory we describe apply to this case. However, assuming randomness is reasonable when the defects are mechanical damage, blemished skins, dryness of fruit, etc., past experience may have convinced the buyer that sub-standard fruit occurs at random.)

The buyer cannot inspect all fruit in many thousands of crates – if he did a lot more might have gone bad by the time he had completed the job! Also, when the tests include cutting the fruit to detect dryness the process would be self-defeating if applied to his whole consignment! He therefore asks a statistician to suggest how many he should sample and what he should decide on the basis of the results.

If the statistician were very busy (or very lazy) he might give an 'off the cuff' answer and say 'Well, take a sample of one hundred, and if you get none or only one sub-standard you'll be pretty safe in assuming the percentage defective in the whole batch is not more than 2 per cent.'

Is this 'off the cuff' advice reasonable? Let's see which of the ideas discussed so far apply here. First, suppose we consider the hypothesis that there are exactly 2 per cent defective in the whole consignment. What is then the probability that we shall find either 0 or 1 defective in a random sample of 100?

In a very large consignment, containing many thousands of grapefruit, the proportion of sub-standard fruit remaining when we take out a relatively small sample does not alter appreciably from 2 per cent and we may reasonably assume our random sample to come from a population with probability $p = 0.02$ of an individual fruit being defective if our hypothesis is correct. We are also assuming, remember, that the state (defective or otherwise) of any one fruit in our sample is independent of the state of any other in the consignment.

Thus we may use the multiplication rule of probabilities to work out the probability that we get any particular sequence of good and bad fruit in our sample. The probabilities of getting any particular total number of sub-standard fruit is in fact given by binomial probabilities of the type met in Chapter 2 (page 44). The probability of a fruit being sub-standard is 0.02; it follows that the probability of the opposite event, namely that a fruit is up to standard, is 0.98. Thus the probability that there are no bad fruits in our sample is $(0.98)^{100}$. One may use logarithms to verify that this has a value (working to 3 decimal places) of 0.133. The binomial probability for exactly one sub-standard grapefruit in the sample can be obtained from a generalization of the formula given in Chapter 2; it turns out to be 0.270. Since the outcomes 0 defective and 1 defective are mutually exclusive we may add the probabilities to get the probability of accepting a consignment with 2 per cent defective or sub-standard. The probability is thus

$$P = 0.133 + 0.270 = 0.403.$$

Thus, if the batch is of the lowest acceptable quality there is only a probability of about 0.4 – or a 40 per cent chance – that we accept it. If the batch contains no sub-standard fruit our scheme is certain to accept it because there would be no defective fruit in the sample; the chance of accepting a perfect batch under this scheme is therefore 100 per cent. At the other extreme, if all the fruit in the sample were sub-standard we reject the consignment because of necessity all the fruit in the sample would also be below standard. What is more interesting is the probability that we accept a consignment with something like 3, 4 or 5 per cent of the fruit below standard. We can use the binomial rules for working out these probabilities. You might try one or two cases for yourself. For example, with 5 per cent of the fruit below standard the value of p is 0.05 and the probability of the sample containing at most one sub-standard fruit is then only about 0.037 or 3.7 per cent.

How would the manager of a wholesale fruit market or his customers feel about this plan? There is only a 40 per cent chance that a batch that is *just* of acceptable quality will be passed as such, while there is a 3.7 per cent chance that a batch with 5 per cent of the fruit below standard will be accepted. To the wholesaler there seems too high a chance of rejecting a marginally acceptable batch – a situation that is of no advantage to the retailer either. The retailer may also be unhappy knowing that if the poor fruit level is as high as 5 per cent he accepts in the long run more than 3 per cent of such consignments.

We can represent graphically the acceptance rate given by our sampling plan for various *true* percentage defectives in a consignment. The graph has on the horizontal axis the true percentage defectives and on the vertical axis a scale representing the probability of accepting a consignment with our sampling scheme. Figure 18 shows the general shape of this *operating characteristic*

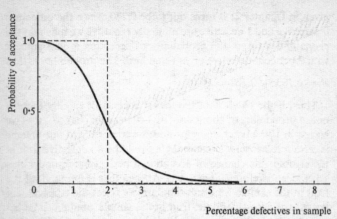

Figure 18 Operating characteristic curve for a sampling inspection
      scheme

curve for a scheme of this sort. Clearly the higher the percentage
of defectives in the consignment the lower the probability that our
sampling scheme will lead to its acceptance. The curve decreases
steadily from 1 to 0 as the consignment or 'population' percentage defectives increases.

Our curve should follow as closely as possible the shape of the
dotted one in Figure 18 under which we accept all batches with 2
per cent or less defective, but cut off sharply to zero at that level.
We aim to get the operating characteristic curve of any sampling
scheme as near as possible to that ideal form.

In a common approach giving us more flexibility we use only
a relatively small sample if it gives strong evidence that our population value is well away from the critical cut-off value of 2 per
cent. If it seems likely to be hovering around the critical 2 per cent
we use a larger sample. One such scheme uses this procedure: we
take a sample of 100; if this contains no defectives we accept the
consignment; if we get more than two defectives we reject the
consignment; if we get either one or two defective we take a
further sample of 100. If the total number of defectives in the two
samples is 3 or less we accept the consignment; otherwise we
reject it. Schemes like these give operating characteristic curves

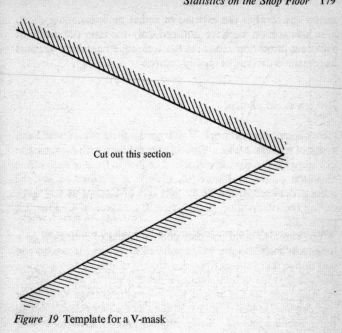

Figure 19 Template for a V-mask

Cut out this section

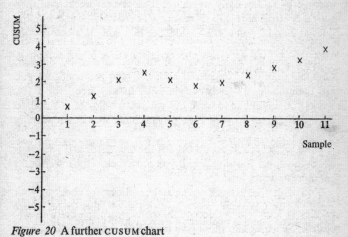

Figure 20 A further CUSUM chart

nearer the ideal at the expense of rather more sampling effort.

In this section we have outlined only the bare rudiments of sampling inspection schemes; the technique has many variants important in the field of quality control.

Are you under control?

Here's some practical work. You'll need a sheet of cardboard and a pair of scissors. Make a V-mask in cardboard with the template in Figure 19. Placing the vertex half an inch in front of each of the deviation points in Figure 20, determine whether or not the process depicted in the figure gets out of control at any stage during the observation period. See if we agree in Chapter 13, pages 230.

You can also build up some control charts of your own on a long-term basis. Do you weigh yourself every day? If you do you will notice fluctuations from day to day of one or two pounds. Assume that the daily weight of a healthy person who is neither 'gaining' nor 'losing' weight has a certain average with inner warning lines at plus or minus 2 pounds on either side of this and outer warning lines at plus or minus 3 pounds. Determine your average by taking daily weighings over several weeks and then start a control chart with this average as central line and warning lines at 2 and 3 pounds each side. See how long it is before your weight gets out of control.

Some people's weights vary from day to day by different amounts than those suggested above. If you are confident you are maintaining a steady weight without any trend for a few weeks you might like to work out your own estimate of standard deviation and set up your own private inner and outer warning lines.

Once the scheme is set up cut out the cream cakes if you overshoot the upper warning line. If you move below the lower limits eat more or worry less!

You might like to work out a CUSUM scheme for checking your weight as well. We can't really give you any further sound advice in Chapter 13 on this problem; it's a real do-it-yourself effort.

10 Picking the Winners

The great selection problem

No, this is not a chapter for gamblers and racing enthusiasts. The statistical selection problem crops up in several areas – the pharmaceutical industry, plant-breeding and industrial chemistry to name but three – where it is relatively easy to produce many thousands of different products of which only a few will have the properties one is seeking.

A plant-breeder may produce thousands of new strains of, say, raspberries, but only a few have properties that make them worth developing for the commercial market. Some sort of test criteria is necessary to single out as soon as possible the most hopeful candidates for development. It takes time to develop a strain by crossing existing varieties and to collect evidence on, for example, the quality of fruit it produces and its likely cropping potential.

One must find out not only whether the fruit will be commercially popular, but whether the plants are disease-resistant, whether they produce adequate crops and also whether the fruit stands up to being transported to market. One can only answer all these questions by growing quite a number of plants of each cross, a process taking several years at best and requiring considerable land, manpower and patience.

Now a breeder who develops perhaps 700 crosses expects less than five or ten of these to have any commercial prospects. It will save both time and money to eliminate many non-starters on the basis of simple experiments that involve little material and labour. He can look first at one of the vital characteristics that determines commercial success. If the new variety is to be grown in an area where a certain disease is prevalent, resistance to that disease may be a prime consideration. If a simple preliminary experiment in the field or the laboratory can test for resistance, all non-resistant varieties can be eliminated immediately.

Where does the statistician come into this? He can deal with random variability or noise. Susceptibility of plants to disease depends upon weather, proximity to sources of infection, cultural methods, etc. and some account needs to be taken of these in planning experiments. Not only may too many plants slip through the net at the early stage, but we may lose potentially useful crosses if we take decisions on the basis of too little evidence. We need a statistical balancing act. Preliminary testing should eliminate non-commercially viable types and pass the good ones for further and more detailed testing. In doing this it should keep our number of 'misclassifications' to a minimum.

The same sort of situation arises in the drug industry, where a pharmaceutical firm may produce thousands of different chemicals while looking for a cure for a disease. Perhaps no more than one in a thousand is likely to pass all the rigorous tests needed before a drug is marketed.

Tests are carried out on animals firstly to assess the effectiveness of the drug for its primary purpose of curing or controlling a disease and secondly to see whether there are any unpleasant or dangerous side-effects. Products that pass these earlier tests may well be subjected to more intensive tests still with animals. Different doses may be used, different periods of treatments and so on, may be tried. Finally, if a product passes all these tests it is tried on humans.

It may take several years' work and cost thousands of pounds to do a complete range of tests on any *one* product. In addition to high costs there are ethical reasons why detailed tests should not be carried out on drugs that are useless or even potentially dangerous either to animals or humans. It is important to screen out drugs that are ineffective or dangerous on the basis of simple preliminary tests.

Schemes for doing this nearly always result in the loss of some potentially good products because noise or random variation results in our making a wrong decision for some products. We also waste time doing further tests on unsatisfactory products that pass through our screening net.

A product that does not achieve the objective of providing a cure is called an *inactive* and one that does is called an *active*. The preliminary tests are *screening tests* and the aim of these – never achieved 100 per cent in practice – is to pass all actives for further testing and to eliminate all inactives. An interesting account of screening procedures for drugs is given by Dunnett (1972) in a book that also gives a description of a number of other important applications of statistics.

Hunting a killer

Let's look at an example that has many features in common with the drug-screening methods described by Dunnett. There are some differences in detail due to the nature of the problem.

The example we consider involves industrial chemistry and the preparation of a product to control an insect pest.

A horticultural pest such as the familiar greenfly often becomes immune to an insecticide. This apparent immunity is really due to genetical selection of resistant strains. It is commonly found that the basic ingredient in the insecticide should be retained with a new additive that can deal with the increasing numbers of resistant flies.

The research chemist probably knows that there is a good chance of finding a suitable additive from a certain class of chemical compound containing as many as 2000 different chemicals; of these only some 40 to 50 are likely to be active.

For simplicity we ignore the possibility – which might be very real in practice – that the additives may actually destroy what effectiveness is still left in the original material to which it is being added.

A suitable preliminary test might be to include each additive with the existing insecticide, which we shall refer to as the *standard*, and to test each new product upon batches of 100 greenfly selected from a large colony of flies. We might test each at a fixed dose level upon four batches and note the number killed in each batch.

The statistical methods for analysing the results of such

screening tests have been developed largely since the war. The tests are designed to enable us to make a clear-cut decision either to reject an additive or to submit it to further tests, in such a way as to minimize errors in classification. Our success in achieving our objective depends upon the way we design our screening tests.

Suppose it is known that although the standard is not doing the job it used to do, there is a dose level that will kill about 50 per cent of the flies. We hope that with an active additive the kill at the same dose level will exceed this. A practical problem is that the toxic effect of a spray varies quite markedly from one batch of greenfly to the next, even if the batches all come from the same colony. If batches come from different colonies the resistances may vary even more markedly. We therefore need replication and also some means of establishing the resistance level of each colony to the standard spray. To establish this *base* level for the standard spray it is usual to take rather more replicates to obtain an accurate base for comparison of toxicity levels for a number of different additives.

If we use four batches of 100 insects each for every additive it is wise to use perhaps six or eight batches of the standard alone.

Because of its unique role in providing a measurement base the standard alone is often referred to as a *control*.

Suppose that for eight replicates of the control the numbers killed per batch of 100 flies are

52, 39, 61, 43, 52, 44, 53, 48,

giving a mean kill per batch of 49.0. For a typical spray with additive one might get for four batches kills of

66, 57, 42, 61,

giving a mean per batch of 56.5.

Do any of the techniques we know at present enable us to decide whether or not this additive is active? Analysis of variance or some extension of the *t*-test to see whether the results indicate a significant difference between 'population' means for the standard and the spray with additive might be a possibility. But remember our warnings in earlier chapters about significance being largely a reflection of a real difference being of a size that our experiment is able to detect it. This approach hardly seems appropriate here, especially as we have so few readings.

We get over this difficulty because a laboratory carrying out screening tests has a pretty good idea what differences are important in commercial terms and also of the order of magnitude of an effect likely if an additive is active. This sort of *prior information* helps in designing test procedures.

In the simplest type of situation a chemist may postulate either that additives are completely inactive or else that they are active to the extent that a given concentration increases the average kill about 15 per 100 above that for the standard dose. This assumption, oversimplified though it may seem, is often fairly realistic in practice. Using it we may formulate hypotheses and study the expected behaviour of various *decision rules* to be associated with our screening experiments.

For example, we may decide to submit to further testing all additives that give a mean kill over the four batches 10 insects greater than the mean kill for controls for that particular colony. Note that we shall use the *one* control of eight batches as a base indicator for *all* additives tested on batches of flies from the same colony.

Looking at the performance

To study how the above rule performs, let's suppose (although we cannot know in advance whether we are correct in doing so) that only 40 of 2000 additives we test are active and, as above, that any active additive gives a population mean kill 15 insects greater than the control. All items that pass the preliminary test, i.e. give a mean kill 10 or more above the control, are subjected to further tests that require ten times the resources of the preliminary tests. The screening is regarded as complete at this stage, and it can be assumed that virtually all inactives have been rejected and only actives go on to more elaborate testing.

We are interested particularly in the number of actives passing the screening test per thousand units of testing effort; this is known as the *yield* of the test procedure.

To calculate the yield we must measure the 'noise' or random variation that leads to differences between responses of batches of 100 to the same insecticide. We could measure this by calculating s^2 based upon observed control values and for replicates of each additive being tried, i.e. one s^2 based on all tests.

Often s^2 is known more accurately from an accumulation of past results for tests on similar compounds and then we may use the value of s or s^2 from these. This value of s is then used to form the standard error appropriate, that for the difference between the mean for eight replicates of the standard and four replicates of the standard plus additive. (We omit details of how it is calculated.)

For illustrative purposes we assume that this standard error has the value 4; in practice it is more likely to be an awkward number, like 4.17 or 5.32.

If an additive is inactive the *population* mean difference between control and standard plus additive will be zero. Using appropriate tables we can, on the assumption that noise has a normal distribution, work out the probabilities of observing a difference greater than 10 between sample means for the standard and the spray with each additive when the true difference is zero. We are only interested in cases where the sample mean with the additive exceeds that for the standard, since a difference in the opposite direction would clearly support more strongly the hypothesis of 'no effect' than one of 'beneficial effect' for the additive.

We can work out the relevant probability for any one sample using a table like Table 35. First we have to *standardize* the difference we are examining. We do this by calculating X given by

$$X = \frac{\text{difference between sample means} - \text{difference between population means}}{\text{standard error of difference between sample means}}.$$

For the inactives the difference between the population means is zero and the standard error of the difference is 4. Our concern is in the difference between the sample means exceeding 10. These numbers give a value for X of 2.5. Now Table 35 tells us that the probability that $X = 2.5$ or some higher value is 0.00621. Now we have assumed that 1960 out of 2000 additives are inactive. Since *each* inactive has a probability of 0.00621 of giving a sample

Table 35 *Probabilities that a standardized normal variable exceeds a value X*

X	Probability	X	Probability	X	Probability
0.00	0.5000	1.00	0.1587	2.00	0.022 75
0.10	0.4602	1.10	0.1357	2.10	0.017 86
0.20	0.4207	1.20	0.1151	2.20	0.013 90
0.30	0.3821	1.30	0.0968	2.30	0.010 72
0.40	0.3446	1.40	0.0808	2.40	0.008 20
0.50	0.3085	1.50	0.0668	2.50	0.006 21
0.60	0.2743	1.60	0.0548	2.60	0.004 66
0.70	0.2420	1.70	0.0446	2.70	0.003 47
0.80	0.2119	1.80	0.0359	2.80	0.002 56
0.90	0.1841	1.90	0.0287	2.90	0.001 87
				3.00	0.001 35

mean difference from control greater than 10, we may calculate the number in 1960 expected to do so as

$$1960 \times 0.00621$$
$$= 12.17,$$

where, as usual, this *expected* number is the theoretical mean or average number. In any particular experiment we expect about this number of inactives to give a difference in excess of 10 – perhaps 11, 14 or 12 would actually do so in any given experiment.

Let's turn to the actives now. For the actives, we suppose in each case that the *true* population difference is 15. An observed difference between the sample means of 10 in this case gives an X value of

$$X = \frac{10 - 15}{4} = -1.25.$$

We have a minor difficulty to overcome before we can use Table 35 to calculate the probability that X exceeds this value for an active. This arises because Table 35 contains no negative values. The required probability is obtained by subtracting the entry for the corresponding positive X from 1. When we do this and look for $X = 1.25$ we run into another snag; there is no such value for X in the table. We can get a good approximation by noting that 1.25 lies midway between 1.20 and 1.30 for which we have entries. The corresponding probabilities are 0.1151 and

0.0968, and for $X = 1.25$ we take a value midway between these two; this is given by the average of the two entries, i.e.

$$\left(\frac{0.1151 + 0.0968}{2} \right)$$
$$= 0.1059.$$

Subtracting this from 1 gives the probability of X exceeding -1.25, i.e.

$$1 - 0.1059 = 0.8941.$$

We have used here a method known as *linear interpolation*; it is useful for getting 'in between' values in tables, although it is usually not completely accurate. More extensive tables show the correct probability in this case to be 0.8944. Thus with 40 actives the expected number to pass at this preliminary stage would be

$$40 \times 0.8944 = 35.8.$$

Note that to this degree of accuracy we get the same expected number if we take the probability to be 0.8941.

The total expected number of actives and inactives to pass the preliminary tests is therefore

$$35.8 + 12.2 = 48.0,$$

providing our assumption about the respective numbers of actives and inactives is correct. Now each additive that passes requires the further tests we specified, equivalent to ten additional first-stage tests. If we regard each first-stage test with 4 replicates of an additive as a 'unit' of testing work, then the expected number of additional test units required to complete screening is

$$10 \times 48.0 = 480.$$

The total number of test units up to the end of the screening stage consists of 2 for controls (because we had 8 replicates), 2000 for testing each of the additives with 4 replicates and 480 second-stage unit equivalents, giving a total of

$$2 + 2000 + 480 = 2482.$$

If we assume that the second-stage tests screen out all the inactives that passed the first test, but retain all the actives that passed it, then the expected number of actives to pass the screening stage is 35.8. On average we lose

$$40 - 35.8 = 4.2$$

actives at this first stage. The *yield*, or number passing per thousand test units, is thus

$$\frac{35.8}{2482} \times 1000 = 14.4.$$

Suppose we had used 10 test units on each of the 2000 additives and so isolated all 40 actives successfully; then we use 20 000 test units (strictly speaking 20 002 if we include controls), and the yield is 2.0, far worse than under our two-stage scheme. The practical cost of obtaining more than a seven-fold increase in yield is to lose some 4 or 5 actives out of 40.

One is left with a slight fear that some wonder substance may be among the 4 or 5 actives that were never fully tested; but such a thing happening is unlikely – a 'wonder' additive would probably do substantially better than an ordinary active. It would pass the test because its performance would reflect a difference from controls even better than the assumed 15 for actives.

Those actives that pass the screening tests are culled further because of undesirable side-effects, difficulty of manufacture, chemical instability and so on. Undesirable side-effects of insecticides that must be looked for are damage to plants, effects on other insects such as bees that are essential for pollinating the crop being grown, toxicity to humans who may eat vegetables or fruit sprayed with the insecticide or toxicity to farm animals. Other dangers include persistence of the chemical in the soil or pollution of rivers and streams and the health hazard to farm workers handling the concentrated chemical or breathing the spray when applying it. It is not unknown for all active additives in a group to be rejected on one or more of these grounds, and then the search starts again with a fresh group of potentially useful substances.

Changing the rules of the game

We've left several important questions unanswered. What happens, for instance, if we were wrong in guessing that the average difference between an active and an inactive was an increase in the 'kill' of 15 for a batch of 100 insects? What if our assumption of 40 actives and 1960 inactives among the 2000 additives is wrong? What if we alter our cut-off for passing or failing at the preliminary test from a sample mean difference from the standard of 10 to, say, 9 or 12? Can yield be improved by some modification of the test procedures?

In discussing some of these points it is helpful to use the terminology *false positive* for an additive that passes the first stage yet is inactive and *false negative* for an active that is rejected after the preliminary test.

Intuitively we see that increasing the difference in means from the control to 12 (instead of 10) before passing an additive decreases the number of false positives; at the same time it increases the expected number of false negatives. Statistical reasoning confirms our intuition, as we shall now see.

Suppose we increase our cut-off difference to 12 and retain our assumption that the standard error of the difference is 4 and that the true difference for an active is 15. We can calculate that for an inactive that gives a false positive $X = \dfrac{12}{4} = 3$. Table 35 then tells us that the probability of a false positive is 0.00135. Thus the expected number of false positives from 1960 inactives is

$$1960 \times 0.00135 = 2.6.$$

For an active to pass, the critical value of X is -0.75. Using linear interpolation we evaluate the relevant probability as

$$1 - \left(\frac{0.2420 + 0.2119}{2} \right) = 0.7731,$$

whence the expected number of actives out of 40 passed for further testing will be

$$40 \times 0.7731 = 30.9.$$

The expected number to be subjected to additional tests is now

$$2.6 + 30.9 = 33.5$$

and the total number of test units to the end of the second stage
has the expected value

$$2 + 2000 + 10 \times 33.5 = 2337.$$

We have reduced the expected number of tests from 2481 in the
last section to 2337, but we have also reduced the yield because
the expected number of actives to pass decreases from 35.7 to
30.9. The yield is now

$$\frac{30.9}{2337} \times 1000 = 13.2,$$

compared to the earlier value of 14.4.

We can work out the yield corresponding to a range of different
cut-off values and find that there is a cut-off point that gives maximum yield. Once the yield has been determined for a few cut-off
values near the optimum, graphical methods may suffice to determine this. The graph has a shape like that in Figure 21.

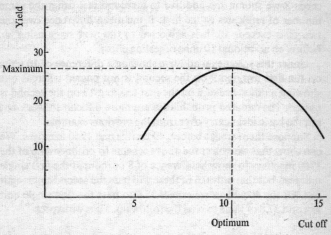

Figure 21 The relationship between cut-off point and yield

What if we have made a mistake in taking the proportion of actives to be 40 in 2000? A strong point of the method is that it can be established mathematically that the cut-off value for difference between means that gives maximum yield does *not* depend on the proportion of actives actually present. For example, suppose the cut-off giving maximum yield with 40 actives and 1960 inactives for some testing scheme is 10.7; it remains 10.7 if there are 70 actives and 1930 inactives. However, the actual value of the maximum yield is different in the two cases.

Things are not so simple if we guess the level of activity of an active incorrectly. If the additives increase the kill not by 15 but by 19 or 21 we require a different cut-off to obtain maximum yield. The best one can do about this is to work out optimum cut-offs for a range of 'active' values that might be of interest and hope to find a compromise cut-off value that gives yields not too far removed from the optima for all of the active levels likely to be of interest. A computer is extremely useful for this task.

It is sometimes possible to increase yield by an initial two-stage testing scheme. In our insecticide example, we might reject after a first test all those items for which the difference in sample mean from the controls is 8 or less. If the difference exceeds 8 we submit the additive to a second test using the same number of replicates as the first. If the mean difference over the two tests exceeds 10, it is subjected to the next stage using, as before, an additional 10 units of testing effort.

Under this scheme an additive showing a difference of 10 units on the first test and 9 on the second is not passed whereas one showing a difference of 9 on the first test and 13 on the second is passed. The required probabilities are more difficult to work out but the basic ideas carry over from the previous example.

Suppose there are, as before, 40 actives and 1960 inactives. We can show that we expect the above scheme to produce 44.6 of the 1960 inactives to show a difference of 8 or more at the first single unit test, but that only 0.6 of these will pass the second single-unit test. Of the 40 actives we expect 38.4 to pass the first single unit test and 37.8 of these to pass the follow up. Thus we expect

$$0.6 + 37.8 = 38.4$$

additives to be subjected to a further 10-unit test. The total number of test units would have expected value

$$2 + 2000 + 44.6 + 38.4 + 38.4 \times 10$$
$$= 2469$$

and the corresponding yield is

$$\frac{37.8}{2469} \times 1000 = 15.3.$$

The increase in yield compared to the simpler scheme in the last section is not dramatic, but we have considered only one two-stage scheme – not necessarily the optimum.

A two-stage scheme often results in a considerable increase in yield. It is possible to increase yield even further by using three-, four-, or even five-stage schemes but the improvement is usually less dramatic and the organizational complexity hardly justified.

Here we have considered a very simple screening problem. We have certainly oversimplified in assuming that additives are either completely inactive or else have a specified level of activity. There may well be a range of active levels which complicates the problem both in theory and practice.

The approach we have used does not suit *all* screening problems. In plant-breeding one may want to select the top twenty varieties from several hundred new experimental strains. The approach to this problem is rather different and theoretically more complicated than the one we have described.

Self service

To check if you have the basic ideas about screening problems take a look at this problem.

Drugs are being tested on dogs to see if they are effective against malignant tumours. A common procedure is to observe after a certain time the size of tumours induced in untreated control dogs and also in dogs treated with the various drugs under test. The mean tumour size for dogs treated with each drug is compared with the mean tumour size for the controls.

Suppose we use 10 control dogs and 5 dogs for testing each drug. An active drug reduces the rate of growth of a tumour so that a smaller mean tumour size indicates an active drug.

Suppose the standard error of the difference between control mean and treated mean tumour weight is 2.0 and that it is decided to pass at the preliminary stage any drug that gives a mean reduction in excess of 4 units. If the mean reduction for an active is 7 units and if all those passing the first stage go through a programme involving 20 times as much in the way of resources as the first stage procedure, what can we say about the yield assuming:

1. that 1000 compounds tested include 50 actives;
2. all inactives are eliminated at the end of the second stage;
3. all actives passing the first stage also pass the second.

Chapter 13, pages 230–31, as usual gives hints.

11   Statistics and the Computer

The influential giant

There can be little doubt about which were the most influential
developments in statistics this century on the practical front. The
first was the development by R. A. Fisher and his co-workers at
Rothamsted of modern experimental design and analysis in the
1920s and 1930s. The second was the widespread introduction of
computers for statistical analyses – a development of the last
twenty years – in which Rothamsted Experimental Station again
played a pioneer role, this time under the leadership of Fisher's
colleague and friend Frank Yates.

The computer has had a strong influence not only upon the way
data are processed but indirectly upon the lines of development of
statistical theory. Much of this influence has been for the good,
but as is almost inevitable in a period of rapid change, there have
also been unfortunate developments.

On the credit side, more elaborate analyses of data can be
carried out at greater speed; it is possible now to carry out
analyses which the time element alone would have precluded had
only non-programable desk calculators been available.

The computer has also provided a valuable tool for simulation –
a topic we touched upon in Chapter 8 when we looked at hospital
appointment systems.

The computer's ability to do elaborate analyses speedily carries
with it a number of pitfalls. For example, we have emphasized the
need to consider what questions are being asked before setting up
the 'model' upon which we are going to base our analysis.
Professional statisticians are all too familiar with the naïve
experimenter who, having used, say, the analysis of variance once,
believes it to be the statistical tool for the analysis of *all* data; if he

has access to a computer he uses one of the many general purpose programs for the analysis of variance when it is quite inappropriate and when he would have done better to seek expert statistical advice. Regression analysis programs in particular get a lot of misuse.

A further danger arises from the widely held belief that a properly programed computer necessarily produces the right answer. There are many pitfalls in numerical work because a computer at best works by approximations. Numbers are stored in the computer's memory banks (and operated upon) only to a certain degree of accuracy. While this is more than sufficient for many business and scientific calculations there are problems for our type of work.

Difficulties also arise if we attempt to use a computer to solve a problem that has no solution – the computer contrives to give us one! We can illustrate this by considering multivariate regression – a topic met in Chapter 7. Let's think first of all about bivariate or two-variable regression and generalize certain ideas from this. It is intuitively obvious that an infinite number of straight lines pass through one point, say (x_1, y_1). We cannot determine the direction of a straight line from the position of one point only on it.

If we are given two or more points lying exactly on the line its direction is determined simply by joining the two points as in Figure 22.

Once we know the position of the line from the two given points on it we may obtain the values of α and β in the equation

$$y = \alpha + \beta x.$$

In regression analysis we have a situation where our points do not lie *exactly* on a straight line. Just as we cannot determine a straight line with only one point, we cannot tell how good an approximation a regression line will be with only two points, for the line of 'best' fit to only two points is the line joining them. We need at least one further point to tell us how good an approximation our least squares line is to a hypothetical 'population' straight line. The more points we have the more evidence do we gain about the 'goodness of fit' of a line to a set of points.

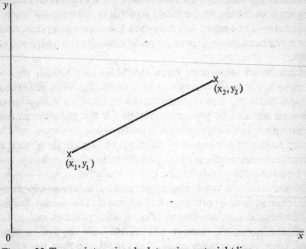

Figure 22 Two points uniquely determine a straight line

Now if there are three variables the function is of this form:

$$y = \alpha + \beta x + \gamma z.$$

This defines a plane in the ordinary three-dimensional space in which we live. Clearly two points in space do not specify a unique plane, for we may take *any* plane containing the line joining the two points and rotate it about this line as an axis to create an infinite number of planes passing through the two points.

If we have three points we may determine a unique plane passing through them *providing* the third point does not lie on the line joining the other two. If points lie only approximately on a plane we shall need at least one additional point, making four altogether, to get some measure of the accuracy with which we are determining the parameters α, β and γ.

Linear equations in more than three variables do not give a geometrical picture, but algebraic analogues can be used. These show a generalization from our result that two points are needed to fix a linear relationship in two dimensions and three points at least in three dimensions: p points are needed, as a minimum, to fix a linear relationship in p dimensions, i.e. a linear relationship between p variables. In the usual regression situation our linear relationship is not exact, so we need at least one extra point, and in practice considerably more, to get information on how good is the fit.

Mathematics looks after these difficulties very well. If we attempt to apply least squares to a problem with p variables when we have less than p observations the theory forces us to a position where we can find no unique estimators for our parameters: the calculation is halted because we have to divide by zero – which cannot be done. In more technical language we are faced with a *matrix inversion* problem which has no satisfactory solution unless we have the required number of observations.

But the computer is not always worried by an impossible problem! It can sometimes proceed when it shouldn't because it rounds off numbers to a certain degree of accuracy – perhaps, say, to 11 significant digits. A number like 217.432 156 983 512 is then recorded in the computer as 217.432 156 98, the four last digits being chopped off; also, a computer sometimes holds a number such as 2.3 in the form 2.299 999 999 9.

After a large number of arithmetic operations, particularly multiplications and divisions, the errors deriving from rounding off build up so much that the computer divides by a small number rather than trying – and failing – to divide by zero when this is what it should be doing. This enables the computer to produce answers to problems that don't have any!

Suppose a social scientist's study, observes, say, 15 variables on each of 10 individuals. An inexperienced worker may try to derive a multiple regression equation in the 15 variables, even though there are only 10 sets of values. A well-behaved computer gives the answer (disguised in some form of computer jargon) 'it can't be done', but many statisticians have seen computers produce apparently respectable – and quite meaningless – solutions, thanks to round-off. Actually at least 15 sets of values, i.e. 5 more individuals, are required, to give a meaningful solution. Even more are required to give any idea of accuracy of fit.

Dodging the wrong answer

There are even simpler situations where round-off may be misleading. This difficulty has sometimes resulted in different formulae to those used on desk calculators being introduced for computers. A simple example is one involving the calculation of the sums of squares of deviations from a mean.

In Chapter 4 we recommend a way of doing this on the back of an envelope or with a desk calculator – namely, first square all the observations and sum these squares, then subtract the square of the sum of the observations divided by the number of observations. This method avoided the heavy computation, sometimes accompanied by round-off errors, that occurs when summing the squares of the actual deviations from the mean, especially when these are ugly decimal fractions.

When using the recommended desk calculation method with large numbers some care is needed to avoid round-off; but it is usually self-evident when it arises on a desk calculator.

However, serious round-off errors on a computer may go undetected unless they produce an answer that is obvious nonsense. This difficulty becomes acute when the sum of squares of deviations from the mean is small but the individual observations are large, so that their sum of squares and the square of their sum are both large. If these numbers exceed the degree of accuracy for our particular computer, perhaps eleven digits, round-off will occur.

Let's look at a simple example and consider the sum of squares of deviations from the mean for the numbers

1 234 567.7, 1 234 567.8, 1 234 567.9.

The mean of these numbers is immediately seen to be 1 234 567.8. The deviations from the mean are respectively −0.1, 0.0 and 0.1, and the sum of squares of these deviations is 0.02. Here direct calculation of the deviations and then squaring them is easier than using the 'desk machine' formula. If we use the desk machine formula we make use of the well-known result we introduced at the end of Chapter 4 (page 94), namely that one may subtract the same quantity from all observations without altering the sum of squares of deviations. By subtracting, for example, 1 234 567.7 our three observations become 0.0, 0.1, 0.2.

However, a computer programed to use the desk calculator formula goes ahead using the original numbers but only recording numbers to 11-digit accuracy at each stage of the calculation.

Table 36 shows what it calculates. The first column specifies the operation that is being performed. The second column gives the exact arithmetic result of the operation and the third column gives the result to 11 significant digits, the supposed machine accuracy. A dash indicates a position in which the computer gives a completely unreliable digit.

Table 36 *Using a desk calculator formula on a computer*

Operation	*True result*	*Computer result*
Square 1 234 567.7	1 524 157 405 883.29	1 524 157 405 9 – –.– –
Square 1 234 567.8	1 524 157 652 796.84	1 524 157 652 8 – –.– –
Square 1 234 567.9	1 524 157 899 710.41	1 524 157 899 7 – –.– –
Sum of squares	4 572 472 958 390.54	4 572 472 958 4 – –.– –
(*Square of sum*) 3	4 572 472 958 390.52	4 572 472 958 4 – –.– –
Difference	0.02	0 – –.– –

From Table 36 it is clear that the method works with all 15 significant digits retained (column 2) but when the computer uses 11 correct digits we effectively get a random answer composed entirely of round-off. On some installations (including ours at the University of Dundee) this example gives a negative value for the sum of squares of deviations – a mathematical impossibility: the square of zero is zero and the square of any other real number, be it positive or negative, is *always* positive, making a negative sum of squares impossible. Yet if the result is wrong and plausible – rather than clearly absurd, as this one would be if negative – round-off errors on a computer may go undetected.

The above example is somewhat artificial and often the desk calculation formula is quite satisfactory on the computer, breaking down only with large numbers or very many observations.

On a computer direct summing of squares of individual deviations from the mean is preferred, for this involves smaller numbers than the original data. Even this method has technical disadvantages and the sum of squares is often calculated using a *recursive* formula. This is a formula that calculates the sum of squares of deviations for $(r + 1)$ observations given the result for r observations and the value of the $(r + 1)$th observation. The method is often used in statistical *packages* – the name we give to collections of ready-made statistical program collections.

Using computer packages

Most computer installations in universities or research institutes and some in industry or business have *program packages* of various types for statistical analyses. They range from collections of programs for analysis of variance, regression and so forth which are implementable only on the particular computer for which they were written, to wide-ranging statistical packages available on hundreds of machines to do many different analyses.

An early example of the latter type of statistical package was the BMD package developed originally at the University of California Los Angeles campus by W. J. Dixon and his co-workers. Perhaps even better known is the SPSS program suite developed in Chicago by a team producing programs for use in analysing social sciences data. At a slightly more specialized level J. A. Nelder and others at Rothamsted Experimental Station have developed a program called GENSTAT primarily for the analysis of experimental and survey data.

BMD is still widely used on both sides of the Atlantic and will be of considerable interest to historians of the subject as the first major example of a statistical package that has proved valuable to many experimenters. Because its many components were written over a period of years by different people in an era when computer development was rapid, the various parts do not all link together smoothly; for example, output from some parts of the package cannot always be used as input to another part.

SPSS was written rather later and deals in particular with the types of analyses that are popular with social scientists. It is more unified than BMD but open to some criticism for it is not as selective as it might be from the statistical viewpoint in guiding experimenters to the right conclusions in any particular problem. There have been several revised versions and an updated SPSS is planned as a joint Anglo-American project using interactive computing.

An *interactive system* is one in which the way the computer is to proceed at each stage is determined by the person running the program and may well depend upon results obtained up to that stage. The interaction is between computer and user of the program.

GENSTAT is at the time of writing less widely available than either BMD or SPSS, but is particularly useful for *data housekeeping* (the sorting-out of data and extraction of information from large bodies of data), the analysis of results using complicated experimental designs and other statistical problems.

In some areas of statistics – particularly regression analysis, there must be nearly as many different program packages as there are computers, and it becomes particularly important to understand the limitations and pitfalls of a particular package. We have already intimated that some regression programs happily produce regression equations where none exist!

Another difficulty with some packages is the vast amount of output they produce (sometimes greater in quantity than the original data) with little or no guidance upon its interpretation. An experimenter can get his data analysed with a program package without any statistical assistance. This is fine if the experimenter knows what he wants and whether his data can answer the questions he is asking by using that particular program. If, however, the wrong program is used on inappropriate (or even on appropriate) data, this prolific output is a waste both of time and expensive computer resources.

There is usually a paucity of information available about what various computer packages accomplish. What there is is usually contained in the *program documentation* supplied with the package. Usually it is reasonably adequate on instructions about how data is to be fed to the computer, what 'control' cards are needed to get a particular form of output, etc., and extremely vague about what the program actually does statistically.

Some recent research in which the present author has been involved has taken place to examine to what extent the computer can take over the role often carried out informally by a statistician when he looks at data, noting the salient features to see what analysis might be appropriate. Computer developments of this type might help 'do-it-yourself' statisticians with little experience in data handling and save them from using inappropriate programs.

A computer can look for 'wild' observations or outliers (which we mentioned in passing in Chapter 4), seek evidence of 'non-normality' in data, or check whether different parts of the data have been recorded with different accuracy. This might happen, for example, if one had population records for different areas. Large towns or cities may have populations given only to the nearest hundred or even thousand, while the populations of smaller towns or villages are given to the nearest ten or even to the nearest unit.

Programs that detect such 'pathological' data and provide, either in the output or the related documentation, advice on what to do with them can help the inexperienced experimenter.

We're not going to go into the technicalities of writing computer programs. Packages are often very complicated technically, and without good documentation even a statistician cannot always be sure how a package will cope with a particular set of data. When a new package becomes available a canny statistician tests it with sets of data for which the analytical results are already available and sees if the results it gives are consistent with already known ones. Ideally, test runs should include examples with peculiarities that will detect any weaknesses in the programming – for example a regression program that produces an answer when the number of observational units is less than the number of variables.

New horizons

We turn now to the more constructive aspects of the computer revolution. In the pre-computer era a multiple regression with four or five variables was all that most people wanted to tackle. Even with a good desk calculator, determining the regression coefficients and confidence limits for a reasonable number of observations could be the best part of one day's work. A computer only takes a couple of minutes to fit a regression with, say, sixteen variables. What is perhaps more important is whether the regression is much use once it has been obtained. We must bear in mind that what a computer *can* do is not always the wisest thing to do. Nevertheless there are many useful statistical techniques that have only been made possible by the computer.

In psychology, for instance, it is common practice to measure a great array of variables, often representing responses of some sort – particularly in studies of learning processes. Techniques were developed early this century to explain thought processes and their workings in terms of relationships between measured variables. The so-called 'factors of the mind' developed by Thurstone (1947) and others gave rise to the modern study, widely used in psychology, of *factor analysis*, a technique that cannot be used on a large scale without a computer. The same technique is now used in other fields – for example by biologists in studying patterns of growth.

The computer has also given us new ways of looking at data. Interactive computing enables us to do the messy arithmetic parts of an analysis quickly and then to decide what to do next in the light of preliminary results. In an analysis of variance, for instance, one can proceed as far as the variance ratio test; if this is not significant we proceed no further. If it is significant a number of treatment comparisons can be looked at individually and various standard errors or confidence limits calculated.

In regression analysis a study of residuals – the differences between observed y_i and their values *estimated* by putting the observed x_i in the regression equation – becomes feasible with a computer; the residuals can give clues on how satisfactory our mathematical model is at describing the data.

Computers and simulation

Perhaps the most novel use of the computer in statistical problems has been in simulation. The use of computers has contributed to developments both in theory and in practice in studies, for example, of stochastic processes.

In Chapter 8 we considered two simulations of a hospital clinic appointment system. Our main interest was in how much time a doctor was idle and how long patients had to wait. The two simulations indicated that there could be appreciable differences.

The computer enables us to carry out a large number of simulations in a short time. One has been used to carry out 100 simulations of the hospital appointment system. For each we have obtained the doctor's idle time, the average patient wait and the number of patients not seen. Although the computer used was by no means the fastest on the market it took less than five seconds to do the 100 simulations. The results are given in Table 37.

Inspection of the table indicates that in the 100 simulations there are 76 occasions when no patient has to make a fresh appointment; on 17 occasions one patient has to make a fresh appointment and on the remaining 7 occasions two have. This implies that in the long run we can reasonably expect all patients to be seen at about three-quarters of the clinics and that only very rarely should we expect more than two patients to have to make fresh appointments. We note that the doctor's idle time varies between 0 and 39 minutes; the average idle time for the doctor over all 100 simulations is 14.44 minutes. This is probably quite acceptable if it is made up of odd one- to five-minute pauses scattered through the morning, giving the doctor reasonable 'breathers'.

The average patient waiting time varies from zero (implying no 15-minute consultations except perhaps for the last patient) to 16.00 minutes. This last average wait occurred on a day when the doctor had no idle time; we can imply that his first consultation lasted 15 minutes and throughout the clinic period he never more than just caught up with the appointment schedule.

Generally speaking one can expect patients' waiting time for later appointments to be longer than for earlier ones. The first patient never has to wait and the second patient's wait cannot

Table 37 *Output of 100 simulations of a hospital appointment system*
$x =$ *Doctor's idle time (minutes)*

$y =$ *Average patient wait (minutes)*

$z =$ *Number of patients not seen*

x	36	20	17	8	0	26	10	19	10	24
y	0.28	4.61	2.00	4.33	12.87	1.44	3.67	1.72	4.44	0.83
z	0	0	0	0	2	0	0	0	0	0

x	28	6	22	2	12	9	28	9	10	0
y	1.83	5.56	0.78	4.56	3.67	2.67	1.06	7.06	5.00	16.00
z	0	0	0	0	0	0	0	1	0	2

x	2	24	4	38	8	0	3	9	7	5
y	8.41	4.28	5.83	0.00	2.28	12.62	6.65	3.06	4.41	10.82
z	1	0	0	0	0	2	1	0	1	1

x	21	0	9	26	27	6	5	13	1	32
y	1.61	15.50	6.71	1.33	1.44	2.94	2.72	1.50	9.53	2.56
z	0	2	1	0	0	0	0	0	1	0

x	2	0	6	24	32	13	3	21	20	8
y	2.89	4.17	5.11	2.72	0.28	2.50	7.72	0.82	1.22	2.50
z	0	0	0	0	0	0	0	1	0	0

x	1	8	0	16	6	14	25	24	39	14
y	6.67	2.28	13.41	4.94	4.11	4.83	0.78	0.83	0.28	5.44
z	0	0	1	0	0	0	0	0	0	0

x	30	26	6	10	34	23	6	2	10	7
y	0.00	0.56	4.33	2.33	0.94	3.50	14.56	6.56	5.33	8.06
z	0	0	0	0	0	0	2	0	0	1

x	25	22	14	9	14	2	26	0	19	12
y	0.78	1.94	2.72	4.33	6.19	5.78	0.94	14.35	1.12	2.06
z	0	0	0	0	2	0	0	1	1	0

x	5	22	30	13	28	18	16	6	28	31
y	6.12	2.67	1.17	3.18	0.50	1.33	1.56	12.53	1.44	0.83
z	1	0	0	1	0	0	0	1	0	0

x	8	2	10	26	26	18	12	6	12	18
y	8.06	5.71	4.00	2.44	1.11	3.00	1.94	13.81	3.39	1.33
z	1	1	0	0	0	0	0	2	0	0

exceed 5 minutes, and then only if the first patient has a 15-minute consultation.

Over the 100 simulations the average wait per patient was 4.25 minutes. In only ten did the average wait for those who actually saw the doctor exceed 10 minutes. The wait for those not seen depends upon when it became clear that the doctor had no chance of seeing them before noon; the patient would then be told immediately, one hopes, to make a fresh appointment.

It is intuitively reasonable that there should be a negative correlation between doctor's idle time and patient's waiting time: if patients wait a long time, the doctor waits little and vice versa.

Hospital administrators could decide if the system we have simulated is satisfactory. Clearly if appointments were set at longer intervals the doctor's idle time would rise but patients would be less likely to wait. Indeed, setting appointments at 15-minute intervals ensures that no patient has to wait. On the other hand the number of patients seen per clinic hour would drop, and unless the total clinic time each day was extended there might be a build-up in waiting lists for appointments.

We mentioned in Chapter 8 (pages 162–3) that simulation studies are often used to compare different systems. To illustrate this let's suppose that the probability of a 5-minute consultation drops to 0.1, that of a 9-minute consultation drops to 0.4 while that of a 15-minute one increases to 0.5. We may carry out a simulation study of the new system with the intervals between appointments remaining at 10 minutes. From carrying out 100 such simulations we obtain a table analogous to Table 37.

We have not reproduced the table here, but it shows that only on 28 occasions of the 100 simulated did no patient have to make a fresh appointment, while on 27 occasions one did, on 28 occasions two did, on 11 occasions three did, and in the remaining 6 simulations four did. The doctor's idle time ranged from 0 to 34 minutes with an average of 4.57 minutes. Only in one case did it exceed 30 minutes and in only 13 cases was it over 10 minutes.

The average waiting time per patient ranged from 1 to 23.07 minutes compared to 0 to 16 minutes in the previous system. The mean, at 9.67 minutes, was more than double that of the previous system.

Clearly the new system is much less satisfactory. In Chapter 8 we noted that the expected consultation time for the earlier system was 10 minutes – the same as the interval between appointments. For the new system the expected consultation time is

$$5 \times 0.1 + 9 \times 0.4 + 15 \times 0.5 = 11.6.$$

This suggests it might be wiser to make appointments at 12-minute intervals.

Many modifications to the general form of an appointment system are possible. Appointments can be made in batches, e.g. 3 at 11 a.m., three more at 11.30 a.m. and so on. This increases the average waiting time per patient, but cuts down the doctor's idle time unless he gets several batches where all patients have short appointments.

Practical simulation studies on the computer are usually much more complicated than those we have described. In studying an industrial process it is often possible to formulate probabilistic models for the flow of raw materials to a plant and for production capacity allowing for routine maintenance, breakdowns, absentee-ism, shortage of materials, etc.

Models can be adapted to study the effects of introducing new machines or altering the size of the labour force or the stockpile level of materials. Studies of this type may require a mathematical model to take into account the relationships between fifty, a hundred or even more factors a great many of which involve an element of probability in their behaviour. Components wear out at a random rate; there is a certain probability that any employee will be absent; one cannot predict the precise arrival times of loads of raw materials on account of traffic delays, vehicle break-downs, etc.

Whether simulations will give an accurate picture depends upon how well the mathematical model reflects real practice. Even in our simple hospital appointment study our simulation would be misleading if we specified consultation times incorrectly.

Simulation is now widely accepted by leading industrial con-cerns as a respectable part of management planning. These studies recently played a vital part in planning a major new steel complex.

Simulations may also help in developing statistical theory. Many theoretical studies lead to complicated mathematical expressions that cannot be evaluated easily. Simulation may be used to study the behaviour of mathematical expressions when we change the values of certain quantities – the parameters – in them. We may for example study the effect of increasing some of these while others are decreased. The method has also been widely used to study robustness of tests such as the *t*-test when assumptions such as that of normality do not hold.

Try simulating again

You had some practice at simulating in Chapter 8. If you want to do some more, look at the effect of altering the interval between appointments from 10 to 12 minutes in the system we introduced on page 207 which had the following probabilities for consultation times: 5 minutes has probability 0.1, 9 minutes has probability 0.4 and 15 minutes has probability 0.5. After your efforts in Chapter 8 you should find it straightforward to do one or two simulations so this time we won't even give any hints in Chapter 13.

12 Guesses and Decisions

What information should we use?

In nearly all the situations examined so far we have based our inferences upon observations in some sort of experimental or investigational set-up. We have had to make certain assumptions *external* to our data when making a test or working out an interval estimate. For example, to validate results based on a *t*-test or to justify confidence intervals involving a value read from a *t*-table we require an assumption of normality.

Sometimes we can justify such assumptions on the basis of the data themselves but assumptions of normality sometimes rest only on a basis of pious hope or, what is a little more satisfactory, on past experience with experiments on similar material.

Suppose I weigh myself on three scales of a particular kind which record the weights 71.3 kg, 72.1 kg and 71.7 kg. This is quite inadequate evidence upon which to decide that the 'errors' in the machines have a normal distribution. If we know that the manufacturers tested 1000 machines and observed that the error distribution had all the characteristics of a normal distribution then we would be much happier to assume approximate normality.

Additional knowledge gained from previous experiments is one way of reinforcing our belief in assumptions. In reality we very seldom do experiments in complete isolation: there is relevant evidence from similar experiments that we or others have done in the past. Can we formally combine information from all these sources in making inferences?

There is a body of theory in statistics on how to combine the results of a number of experiments. This area bristles with practical difficulties because more often than not experiments differ in structure – the treatments included are not the same, the experimental designs are different, the 'noise' or error components are of very different magnitudes, external conditions (e.g. weather conditions from season to season in an agricultural context) differ.

All these problems have been coped with to some extent by extending orthodox theories and modelling processes of the types considered earlier. In discussing sampling inspection plans in Chapter 9 (pages 165–6) we pointed out that our estimate of s^2 was often based on earlier observations and also in Chapter 10 (page 186) we used a similar estimate based on past experience.

A much more diffuse sort of past experience permeates our everyday life – the prior beliefs held with varying degrees of conviction by different people. Can we, or should we, seek a way of taking these into account in our analysis and interpretation of experimental results?

Statisticians tend to fall into two schools in their reactions to this. Some only permit information obtained from an experiment using orthodox methods of analysis of the kind described in earlier chapters and insist on careful specification of what assumptions are made in formulating the mathematical model upon which the analysis is based. Theory essentially depends on probabilities of the kind so far considered in this book, namely probabilities that in essence specify the frequencies in the long run with which certain events will occur.

The second school of statisticians believes that it is quite permissible to attach a probability measure to one's *degree of belief* in some proposition, and indeed this is what we do in everyday life. This concept would be extremely useful if all men with the same prior knowledge arrived at the same value for a probability. Many people believe that all *rational* men given the same prior information should have the same degree of belief in a proposition. For example, if the only source of information were the orthodox version of the Bible they believe that all rational men should, after reading the Bible, attach the same probability measure to the belief 'God is a Tory'. The difficulties here are immense. How does one define a rational man (or for that matter a Tory?) It is likely that the Archbishop of Canterbury and the Prime Minister of Great Britain might put different degrees of belief on this proposition, yet who is to say which is more rational?

However, those who think we should take our degrees of belief into account have something on their side. In deciding whether or not to take an umbrella when we leave home for work we make our own assessment of a number of factors; last night's TV weather forecast, this morning's radio forecast, what it looks like out of the window, whether our chilblains itch, whether the cat wants to go out.

We all have different 'weights' which we put on these indicators. Most of us put more weight on this morning's radio forecast than on last night's TV forecast, because it is more up to date. Some ignore the cat's behaviour whereas others consider it a more reliable weather indicator than the official forecast – there is, after all, a lot of evidence that animals are sensitive to impending weather changes. Also, weather is sometimes a very local matter and an official forecast of 'isolated scattered showers' for an area the size of Scotland or New South Wales is not very helpful.

Statisticians who believe in taking prior belief into account argue this way: people have different amounts of prior information and interpret it differently, but this does not matter if they state what these *prior* beliefs are and follow the same rules for combining their prior beliefs with the experimental evidence to produce *posterior* beliefs. This attitude is honest but many think it is also misguided. There remain thorny problems, especially about how one quantifies complete ignorance, many of them in a sense philosophical.

A more practical problem is how much weight one should give to prior opinions – especially if they are vague – and how much weight to hard experimental fact. This dilemma is usually solved by adopting the *Bayesian* approach, so called because it involves a celebrated theorem formulated by the Reverend Thomas Bayes (1702–61) and bearing his name. The theorem as applied to probabilities with a frequency base is not disputed. The controversy rages around the question of whether its application to probabilities that represent personal degrees of belief in particular propositions is meaningful.

Bayes's theorem

Let's first consider non-controversial aspects; we shall neither formally state nor prove the theorem but content ourselves with illustrating how it works for frequency style probabilities by an example. The theorem shows how the probability we initially associate with an event, the *prior* probability, may be modified by additional knowledge. This is used to calculate the *posterior* probability which takes into account the additional knowledge.

Suppose we know that our grocer Mr Bloggs gets $\frac{9}{10}$ of his eggs from farmer Brown and the remaining $\frac{1}{10}$ from farmer Sly. Our best prior estimates of the probability that our breakfast egg comes from farmer Brown is 0.9 and that it comes from farmer Sly is 0.1 – reflecting the proportions from each source. Suppose now the egg we get for breakfast is bad. This information alone is no help in determining the probable source of the egg. But suppose we have somehow acquired the knowledge that eggs supplied by farmer Brown only include 1 bad egg in 1000 while eggs supplied by farmer Sly contain 1 bad egg in 4. Intuitively we see immediately that the egg being bad makes it more likely that it was supplied by farmer Sly.

We can work out the new probability, that is the posterior probability, that the egg came from farmer Sly given that it was bad. Look at it this way. Suppose the grocer's stock of 10,000 eggs is divided in the proportions in which he buys them – 9000 from farmer Brown and 1000 from farmer Sly. If the bad eggs in the stock are also proportional to their probabilities there are 9 bad eggs among those from farmer Brown and 250 among those from farmer Sly, a total of 259. Thus if we get a bad egg the probability that it came from farmer Sly is $\frac{250}{259}$. Thus the knowledge that the egg is bad makes the probability it came from farmer Sly $\frac{250}{259}$ rather than $\frac{1}{10}$, the probability we calculate without this knowledge.

We would get this result by using Bayes's theorem and its use in this context is acceptable to all statisticians. Problems arise when we have no hard facts upon which to base our application of the theorem. People who use Bayes's theorem as a basis for inference have attempted to introduce prior personal or subjective probabilities to indicate their degree of belief in certain hypotheses or to assign probabilities to parameters taking certain values.

Let's take a very simple example. Suppose we are introduced to a man and asked to determine whether he weighs more than 75 kg, or weighs 75 kg or less. By looking at a man and using our general knowledge of the size and shape of human beings, we can form some sort of opinion as to whether he weighed more than 75 kg or not. We might be willing to put a degree of belief on our feelings about his weight. A doctor used to examining patients and assessing their weight might say 'I am 99 per cent certain the man weighs more than 75 kg.' A layman, less used to judging weights of people, might say he was only 80 per cent certain the man weighed more than 75 kg. A person who felt himself a hopeless judge of weight might say he was completely uncertain and felt the man was equally likely to be over or under 75 kg. The degrees of belief of these three persons in the proposition that his weight exceeds 75 kg can be represented by prior personal or subjective probabilities of 0.99, 0.80 and 0.50 respectively.

Suppose we are provided with five weighing machines and told that tests have shown that these machines have come from a large batch which give the correct weight on average but that individual machines have a 'noise' component with a normal distribution. Let us now weigh our man on these five machines. Suppose the recorded weights are 74.2, 74.5, 74.7, 74.9, 75.2. What are the intuitive reactions of our three observers? The doctor is inclined to say 'only the last machine can possibly be correct'. He feels that we have been unlucky with the other four – they must almost certainly be recording under weight.

The man with the prior probability of 0.80 is more inclined to waver and perhaps say 'Well, I guess the evidence is going against me.' The man who just didn't know feels that on the whole the machines were indicating a weight below 75 kg and this is the only real evidence he has. He accepts the evidence the experiment provides – and statistically that evidence is exactly the evidence provided by a standard non-Bayesian analysis of the recorded weights. Using the methods of Chapter 4 we find for instance (you might like to check this for yourself) that 95 per cent confidence limits for the true weight are 74.22 to 75.18.

These limits do not preclude the possibility that the man's weight is above 75 kg but they show quite reasonable grounds for supposing it to be below.

The Bayesian would advocate that the experimental evidence must be combined with the prior probabilities to form *posterior* probabilities. These, in turn, can be regarded as prior probabilities for further experiments. They argue that Bayes's theorem is the correct way of deciding how to give weights to prior information and experimental evidence when combining them. In nearly all practical situations the mathematics of the combining procedure is complicated. It is possible to work out analogues of the confidence limits we obtained before based upon prior beliefs and experimental information.

To the Bayesian the classical approach which uses the evidence of the experiment alone is only justified if we start from complete ignorance. When we have only two hypotheses as we did above prior ignorance can be expressed as an equal degree of belief *a priori* in the two propositions. Extending these considerations to an estimation problem with no prior ideas about the value a parameter may take – perhaps we believe only that its value lies between plus infinity and minus infinity or between zero and infinity – lands us in deep philosophical difficulties. Bayesians disagree amongst themselves about how to deal with this problem and often resort to the methods described in earlier chapters.

The greatest practical disadvantage of Bayesian methods is that different prior beliefs lead to different solutions to inference problems. Although the method is completely honest providing one's prior beliefs are clearly declared the outcomes are of value only to somebody who accepts the reasonableness of the prior probabilities used. This criticism, serious as it is, is somewhat tempered by the fact that different prior assumptions very often lead to virtually the same inferences; often 'classes' of prior beliefs with this fortunate property can be found that satisfy most rational men.

Finally, we must be clear that all of us make use of prior information in some way. We do it when we assume normality purely on the basis of past experience in similar situations. We do it when our past experience tells us that we should 'refuse to believe' the result of an experiment that indicates significance and feel that a 1 in 20 or 1 in 100 chance *has* come up in our particular experiment. In this situation we may ignore the result of our experiment and choose to perform another one.

So everybody uses prior information subjectively. Bayesians make formal recognition of this fact and attempt to put it into a personal or subjective probability framework.

One of the weakest planks in their platform is that Bayesian analyses of data are usually more complicated than classical methods. At the same time their analyses seldom show any real advantage.

A rather different situation prevails in the field of *decision theory* and *decision problems*.

Decision theory

The aim in many examples in earlier chapters has been to reach a decision. Was Dr Gesser really a Guesser? Was there a real difference in response times between left and right visual fields? When should we stop a process because it seems to be out of control? We saw in Chapter 9 that one approach, e.g. CUSUM, might lead us to a decision quicker than another.

We often must make decisions under uncertainty and an important new theory has been developed, again in the last twenty-five years or so, to deal with the problem. We seek a result that is in some sense most advantageous to us. The difficulty is that in the face of uncertainty a decision in any one particular case may be disadvantageous: our aim should be to make our decisions as advantageous as possible in the long run.

Suppose a doctor cannot on the basis of any simple examination tell which of two diseases a patient has because their symptoms are so alike. Now the treatment for these two diseases may be very different and giving the treatment for one disease to a patient suffering from the other may make him worse rather than better. How can the doctor make a rational decision about how

to treat a patient once he has decided a patient needs treating –
but, remember, he is not certain which of two treatments is
needed? Some scale of preferences is needed for the four possible
actions involving the two diseases, which we shall call collywobbles and scratchrash. The possible actions are:

Treat for collywobbles when patient has collywobbles;
Treat for collywobbles when patient has scratchrash;
Treat for scratchrash when patient has scratchrash;
Treat for scratchrash when patient has collywobbles.

The first and third actions are clearly desirable.

The consequences of prescribing the collywobbles cure to a
patient with scratchrash are far worse than prescribing the scratchrash cure to a patient with collywobbles.

For illustrative purposes we suppose that our doctor is a very
mercenary man who is only interested in making as much money
as he can from his patients. Suppose he thinks in this way. If he
cures a patient of collywobbles that patient is extremely happy,
and will pay a bill for £20. If he cures a patient of scratchrash that
patient is reasonably happy, but only happy enough to pay £15
without complaining. If he gives a patient with scratchrash the
collywobbles treatment the consequences are bad and the doctor
can make no charge. Finally, if he gives a patient with collywobbles the scratchrash treatment the patient takes longer to
recover but nothing drastic happens and the doctor can submit a
bill for £3. We set out the gains corresponding to the doctor's
possible decisions for each possible state of nature in Table 38.

Table 38 *Decisions and doctor's gains*

	State of nature	
	Patient has collywobbles	Patient has scratchrash
Action Treat for collywobbles	20	0
Treat for scratchrash	3	15

Our unethical doctor, interested only in money, wants to adopt
the strategy that will maximize his profit. Before he can work this

out he needs some information about the relative frequency with which he is likely to encounter patients suffering from each complaint. Suppose he knows from past experience that the probability of a patient having collywobbles is $\frac{2}{3}$ and of his having scratchrash is $\frac{1}{3}$. (We exclude the possibility of a patient having both diseases at once or any other disease.)

Suppose the doctor decides that he will always prescribe for collywobbles since it is the more common. What then is his expected gain? In the long run on $\frac{2}{3}$ of all occasions he is correct and gets £20. The remaining $\frac{1}{3}$ he is wrong and gets nothing. Thus his expected gain per patient is

$$\tfrac{2}{3} \times 20 + \tfrac{1}{3} \times 0 = £13.33.$$

If on the other hand he treats everybody for scratchrash he makes £15 when correct but only £3 when he makes a wrong diagnosis. His expected gain per patient is then

$$\tfrac{1}{3} \times 15 + \tfrac{2}{3} \times 3 = £7.$$

Suppose now that the doctor decides he is too generous. Instead of charging £3 for treating a patient for scratchrash when he has collywobbles he now asks £15. He now charges £18 when he correctly treats for scratchrash. If he always treats patients for collywobbles his expected gain is unaltered. If he always treats for scratchrash his expected gain becomes

$$\tfrac{1}{3} \times 18 + \tfrac{2}{3} \times 15 = £16.$$

Thus, although his maximum possible gain *in any one case* is still £20 for correct treatment of collywobbles, he makes the greatest profit in the long run if he always prescribes for scratchrash, averaging £16 per patient. Ethically this is better also because prescribing the scratchrash cure to a collywobbles sufferer does less harm than the reverse.

Note that the optimum decision depends not only upon the fees chargeable in the various situations, but also upon the probabilities of each patient having collywobbles or scratchrash. Suppose we adhere to the scale of fees in Table 38, but that the probability of a patient having collywobbles is 0.1 and of his having scratchrash is 0.9. Then if the strategy is to treat all patients for collywobbles the expected gain is

$$0.1 \times 20 + 0.9 \times 0 = £2.00,$$

while if he always treats for scratchrash it is

$$0.9 \times 15 + 0.1 \times 3 = £13.80.$$

It's clear from this example that to make a rational decision about how to maximize expected gain we must have an assessment of probabilities representing the uncertainty and a table of gains for the various situations that might arise. Decision theory brings together two concepts, uncertainty and a scale of values for certain actions, to point to a sensible strategy.

When gains or losses can be expressed in financial terms this may be relatively straightforward, but in many situations 'loss' or 'gain' refers to less tangible concepts such as social amenity. The difficulty then is to place a *utility* value upon such ideas.

The statistical approach to decision-making under uncertainty gives rise to *decision theory*. Many decisions in the face of uncertainty carry an element of *risk*. When risks can be expressed in financial terms they provide a useful basis upon which to quantify.

Quantification is harder when a decision lies between increasing costs and a somewhat unclear social benefit. If an industrial process gives off sulphur dioxide fumes that are annoying to people living near a factory, how much are consumers of a product prepared to pay to have this nuisance eliminated? This may depend upon how badly they want the product; it may also depend upon their social consciences or upon whether they are convinced that the fumes are dangerous to health (as distinct from merely annoying). If the additional costs are such that the demand for the product falls and people lose their jobs at the factory, those irritated by the fumes may be more readily prepared to put up with irritation or even a minor health hazard.

A useful concept in decision theory is that of a *loss function*. This has the general characteristic that the further our decision is at variance with the true state of nature the greater the value of this function. Now decision processes may have various desirable aims. One sometimes seeks a decision procedure to minimize the maximum expected loss: the *minimax* principle. The loss function is relatively easy to specify if loss can be measured in financial terms. If losses are less tangible, such as harm to health or the environment, a subjective element enters into its specification.

An important feature of decision theory is that it usually emphasizes *expectation* rather than the probabilistic aspects of inference, which is where the emphasis lies in hypothesis testing and estimation; one aims to do something like minimizing *expected* loss, or what is equivalent, maximizing *expected* gain when applying decision theory.

In one sense decision theory carries the role of the statistician rather further than the classical theories of inference about parameters, and the subject has many ramifications. If we introduce subjective prior probabilities we have *Bayesian decision theory*.

Some interesting applications of the ideas have arisen with attempts to formalize the way in which doctors make diagnoses. Informally, diagnosis consists of a doctor observing symptoms and using his judgement to assess the probabilities of each as an indicator of a particular disease. He then weighs, again using his own judgement, these various pieces of evidence and arrives at a diagnosis.

In diagnosis he will build in, subconsciously perhaps, some sort of loss function in this sense: if he arrives at a conclusion that the patient has either disease *A* or disease *B*, but he is not sure which, his course of action will depend upon the consequences of a wrong decision. Suppose that condition *A* kills if an operation is not performed within 24 hours, but condition *B* is cured by a week's rest. He must then decide whether or not to submit the patient to a perhaps needless operation or risk his death because he does not operate.

Different doctors may make different decisions in a case like this either because they weigh their evidence differently in arriving at an overall assessment of the likelihood of the disease being *A* or *B*, or because they employ different loss functions in deciding upon the appropriate action to take.

In serious situations most of us prefer an experienced doctor to use his judgement in a life-and-death situation. Much routine medical diagnosis is far less complicated but nevertheless time-consuming. This has given rise to an interest in *computer diagnosis* in routine situations. In this system a computer digests large amounts of information about a patient such as temperature, blood pressure, the results of laboratory tests on samples of blood, sputum and urine. It compares the results with levels known to indicate certain conditions and, if the evidence is clear-cut, diagnoses some specific disease.

All this has a ring of 1984 about it. The mechanics of computer diagnosis could be the subject of a book on its own. However, computer diagnosis is not likely to bring a threat of redundancy to doctors: at best it can reduce their routine work load just as the pocket calculator has taken away drudgery with figures for shopkeepers or office clerks in working out discounts, taxes at a specified rate, and so forth.

In the foreseeable future a doctor will have to make full use of his general experience to confirm computer-based diagnoses; meanwhile the computer will no doubt give rise to many a laugh when it incorrectly diagnoses pregnancy in a patient because it has not been told that the patient is a male.

Decision theory and your pay

Suppose you are a wage negotiator for your trade union and the management side offer you a choice of several pay schemes. Suppose each man can turn out between 80 and 120 units per day, but the more he produces the more likely he is to have a high proportion of faulty units in his output. How would you decide between these three pay offers:

1. a flat rate of £5 per day;
2. a flat rate of 5p per unit of production that is fault free, but nothing for a faulty unit;
3. a flat rate of 7p per unit of production that is fault free, but a deduction of 12p per unit for each faulty unit.

(Only one scheme is to be chosen for all employees.)

Clearly for a man who produces 80 units, all fault free, scheme 3 is the best. If he produces 120 units of which 30 are faulty scheme 1 is best. If he produces 120 units of which 18 are faulty scheme 2 is best.

As a trade union negotiator you have a problem here. You should ask the management side for output records for your members. If they can give detailed records of daily production of good and faulty units for each member over an extended period then it is possible – assuming the same production pattern continues – to work out the scheme that gives the maximum expected daily total taken over all members. This may not be the best scheme, of course, for some individuals.

Problems of this kind – and indeed many other problems of decision making under uncertainty – are discussed in considerable detail and without unnecessary mathematics by Aitchison (1970) in a book entitled *Choice against Chance*.

What's yours?

Suppose the barmaid at your favourite pub is a little forgetful but so friendly and helpful that you don't like to grumble. When you go into the pub you invariably order either malt whisky or blended whisky. Now, you don't like soda in your whisky – especially if it's malt – but you note on your frequent visits to the bar that on one-third of your visits she splashes soda into the drink you order. Your preference for drinks measured in units of euphoria are given in Table 39.

Table 39 *Alcoholic euphoria*

	Malt whisky	Blended whisky
Straight	21	15
With soda	4	12

If you decide to always place the same order, should you order malt or blended whisky? If the barmaid becomes less absent-minded and only puts in soda once per five visits should you change your order?

Next chapter for hints if you are in difficulties and need a cure for a hangover.

13 First Aid Department

Here are the hints if you had any trouble with the examples at the ends of chapters.

Self diagnosis (Chapter 2, page 47)

The first thing we need is the probability that the doctor selects the right patient at a trial if he is just guessing. There is 1 chance in 3 that he will do so, thus $p = \frac{1}{3}$ under the guessing hypothesis.

Using the multiplication rule for independent events, the probability that by guessing he will pick the correct patient in all 5 trials is

$$\tfrac{1}{3} \times \tfrac{1}{3} \times \tfrac{1}{3} \times \tfrac{1}{3} \times \tfrac{1}{3} = \tfrac{1}{243}$$

and this is the size of the critical region.

If he makes no mistake or just 1 mistake in the 5 trials we have to call in the binomial distribution and use it to compute the probability of getting all successes or at most 1 failure when the probability of success at each trial is $\frac{1}{3}$. The probability of 5 successes is, as above, $\frac{1}{243}$. By analogy with the result on page 45, the probability of 4 successes and 1 failure is

$$5 \times \tfrac{1}{3} \times \tfrac{1}{3} \times \tfrac{1}{3} \times \tfrac{1}{3} \times \tfrac{2}{3}$$

since $\frac{2}{3}$ is the probability of failure, i.e., an incorrect guess, and there are 5 mutually exclusive orderings of 4 successes and 1 failure. This probability is $\frac{10}{243}$, thus the size of the critical region is now

$$\tfrac{1}{243} + \tfrac{10}{243} = \tfrac{11}{243}.$$

If the probability that the doctor makes the correct selection is

$\frac{3}{4}$ we do not reject the hypothesis that he is guessing unless he makes the correct selection in all 5 cases. When $p = \frac{3}{4}$ the probability of his doing this is

$$\frac{3}{4} \times \frac{3}{4} \times \frac{3}{4} \times \frac{3}{4} \times \frac{3}{4} = \frac{243}{1024}.$$

The opposite event – namely that he does not select the correct patient at all five trials – has probability

$$1 - \frac{243}{1024} = \frac{781}{1024}.$$

Thus there is a 76 per cent chance that the hypothesis of guessing will not be rejected even when $p = \frac{3}{4}$.

Do you think this is a satisfactory experiment? How does its performance compare with the other experiments we looked at in Chapter 2?

A do-it-yourself chi-squared example (Chapter 3, page 66)

We need the row and column totals in Table 8 to get the expected numbers in each cell. All of these are given in Table 40.

Table 40 *Expected numbers of bananas in each grade*

		Grade		
Shipment	*A*	*B*	*C*	*Row total*
1	20	67	13	100
2	20	67	13	100
3	20	67	13	100
4	20	67	13	100
Column total	80	268	52	400

Note the simple form of the table. All the expected numbers in any one column are the same, because the sample size for each shipment is the same.

The chi-squared statistic is

$$\frac{2^2 + 9^2 + 6^2 + 5^2}{20} + \frac{2^2 + 5^2 + 3^2 + 4^2}{67} + \frac{0^2 + 4^2 + 3^2 + 1^2}{13}$$
$$= 10.1.$$

What about degrees of freedom? The argument on page 56 gives the answer as 6. Table 5 tells us that with 6 degrees of freedom the minimum value of chi-squared for significance at the 5 per cent level is 12.59. Thus we do not reject the hypothesis of different proportions in the various grades for the several shipments. Remember this does *not* prove the proportions are the same.

Some calculations to try (Chapter 4, page 93)

Adopting the suggestion to subtract 285 from all observations the data become

$$-3, 12, -2, 1, 6, 9, -2.$$

As indicated on page 94, this has a mean of 3, corresponding to an original data mean of 288. To get s^2 we first square and add the above numbers, getting 279. The total of the numbers is 21. Squaring this and dividing by the number of observations we get

$$\frac{(21)^2}{7} = 63,$$

whence

$$s^2 = \frac{279 - 63}{6} = \frac{216}{6} = 36.$$

To apply the formula on page 91 for confidence limits we have $n = 7$, $f = 6$, $\alpha = 0.1$, $\bar{x} = 288$. Table 11 shows the appropriate value of $t_{f,\alpha}$ to be 5.96. Therefore the appropriate confidence limits are

$$288 \pm 5.96 \times \frac{6}{\sqrt{7}}$$

$$= 288 \pm 13.52.$$

To verify our assertions about the effects on mean and s^2 of subtracting the same thing from all observations, suppose that from each x_i ($i = 1, 2, \ldots, n$) we subtract a constant a. We then get a 'new' set of observations for which we write

$$y_i = x_i - a.$$

Now, for their mean we have

$$\bar{y} = \frac{y_1 + y_2 + \ldots + y_n}{n}$$

$$= \frac{(x_1 - a) + (x_2 - a) + \ldots + (x_n - a)}{n}$$

$$= \frac{x_1 + x_2 + \ldots + x_n - na}{n}$$

$$= \frac{x_1 + x_2 + \ldots + x_n}{n} - a$$

$$= \bar{x} - a,$$

as we asserted.

If we write

$$s_x^2 = \frac{\Sigma(x_i - \bar{x})^2}{n - 1}$$

(the thing we previously called s^2)
and

$$s_y^2 = \frac{\Sigma(y_i - \bar{y})^2}{n - 1}$$

then we easily verify that $s_x^2 = s_y^2$ since

$$y_i - \bar{y} = x_i - a - (\bar{x} - a) = x_i - \bar{x}.$$

Try one yourself (Chapter 5, page 114)

The experiment involves a randomized blocks design with brands of petrol corresponding to treatments and types of car to blocks.

Subtracting 40 from each entry in Table 22 gives the values in Table 41 where row and column totals have been appended.

Table 41 *Subtraction of 40 from Table 22 data*

Brand	1	Type 2	3	4	Brand total
A	2	−3	9	15	23
B	−1	−2	7	13	17
C	1	−2	9	12	20
Type totals	2	−7	25	40	60

We've had quite a bit of practice now at getting sums of squares of deviations, so we shall proceed straight to the analysis of variance set out in Table 42.

Table 42 *Analysis of variance of petrol data*

	Degrees of freedom	Sum of squares	Mean squares	Variance ratio
Between types	3	459.33		
Between brands	2	4.50	2.25	1.65
Error	6	8.17	1.36	
Total	11	472.00		

The variance ratio of 1.65 is well below the *F*-value required for significance. We can easily see from Table 19 that the minimum value required for significance is about 5. Why?

There may be little point in working out confidence limits for differences between means in this situation. But note that they are not affected by our subtracting 40 from all observations, as this does not alter either the difference between two means or the sums of squares of deviations from the mean.

Keeping yourself amused (Chapter 6, page 122)

A fairly easy example this time. The expected number to answer the first question is

$$\tfrac{1}{3} \times 135 = 45.$$

Of these, again one third should have an age divisible by three (because one third of all integers are divisible by three), so we expect 15 'yes' answers to the first question. This leaves

$$72 - 15 = 57$$

as the expected number of 'yes' answers from the 90 expected to answer this question.

Now 57 out of 90 is

$$\tfrac{57}{90} \times 100 = 63.3 \text{ per cent.}$$

Easy, isn't it?

How did the studies of literary style go? We can't give you much help here because we don't know what you looked at.

For your own calculations (Chapter 7, page 148)

We gave you the sum of the squares of x and the sum of the products of corresponding x and y values. All we have to do now is work out the 'corrections' to be subtracted from these, i.e.

$$\frac{(\Sigma x_i)^2}{n} \text{ and } \frac{\Sigma x_i \Sigma y_i}{n}$$

and we are well on the way to getting a and b. Now $\Sigma x_i = 162$ and $\Sigma y_i = 329$ and $n = 10$, whence

$$\frac{(\Sigma x_i)^2}{n} = 2624.4 \text{ and } \frac{\Sigma x_i \Sigma y_i}{n} = 5329.8.$$

Subtracting these from the sums of squares and products given on page 149 we get

$$\Sigma X_i^2 = 797.6 \text{ and } \Sigma X_i Y_i = 1602.2$$

whence

$$b = \frac{\Sigma X_i Y_i}{\Sigma X_i^2} = \frac{1602.2}{797.6} = 2.009.$$

Also $\bar{x} = 16.2$ and $\bar{y} = 32.9$ whence

$$a = \bar{y} - b\bar{x}$$
$$= 32.9 - (2.009 \times 16.2)$$
$$= 0.354.$$

Thus the regression equation is

$$y = 0.354 + 2.009x.$$

You should plot this line on a graph together with the data to see whether the fit looks reasonable. For 22 days' storage we get our estimated number of rotten oranges by 'plugging in' $x = 22$ in the regression equation, giving

$$y = 0.354 + 2.009 \times 22$$
$$= 44.55.$$

After 60 days the estimate for y is 120.89.

Not too many reservations are needed about the first estimate, but the second one must be treated with caution because it is well beyond our range of observations. There might not be 120 oranges in a crate! Even if there were there would be some doubt about the trend continuing linearly for as long as two months. After a certain time oranges are likely to reach their keeping limit and the rate of rotting is likely to increase, destroying the linear relationship.

Self simulation (Chapter 8, page 163)

To get the correct probabilities for a consultation being of a particular length, associate a random digit with each patient. If that digit is 0 or 1 assign a 5-minute consultation (there is a probability of $\frac{1}{5}$ of getting these two digits at any point in a random sequence). Similarly if 2 or 3 occur assign a 6-minute consultation; if 4, 5, 6 or 7 occur assign a 12-minute consultation; and if 8 or 9 occur assign a 20-minute consultation.

We could allocate the random digits to appointment times in other ways; for example 0, 1 to a 20-minute consultation; 2, 3, 4, 5 to one of 12 minutes; 6, 7 to one of 6 minutes; and 8, 9 to one of 5 minutes.

The expected consultation time is found by multiplying each time by the probability a consultation lasts that long and adding the results to get

$$E = \tfrac{1}{5} \times 5 + \tfrac{1}{5} \times 6 + \tfrac{2}{5} \times 12 + \tfrac{1}{5} \times 20 = 11.0.$$

Since the average consultation time exceeds the interval of 10 minutes between arrivals we might expect queues and patients to have noticeable waits. Does your simulation indicate this? Would it be better to increase intervals between appointments slightly?

Suppose the system is modified so that there is a probability of 0.1 that a patient does not turn up. The simulation can cope with this by assigning two random numbers to each patient who turns up. If the first one is zero (with probability $\frac{1}{10}$) the patient should be regarded as not turning up and no further action is needed. Otherwise a second number is selected to assign consultation time.

Are you under control? (Chapter 9, page 180)

Well, things didn't get out of control. Agreed?

Self service (Chapter 10, page 193)

First we want to know the probability that an inactive will pass. If we calculate X (page 186) for an inactive, taking the difference between sample means as 4, the difference between population means as zero, and the relevant standard error as 2 we get

$$X = \tfrac{4}{2} = 2,$$

and from Table 35 we read off the relevant probability as 0.022 75. Thus from 950 inactives we expect

$$0.022\ 75 \times 950 = 21.6$$

to pass.

For the actives with a population difference of 7 the critical X value is

$$X = \left(\frac{4-7}{2}\right) = -1.5,$$

and using Table 35 in the way described in Chapter 10 the corresponding probability is

$$1 - 0.0668 = 0.9332.$$

Thus from the 50 actives the expected number to pass is 46.7.

The expected number of compounds to be submitted to 20 additional test units each is thus

$$21.6 + 46.7 = 68.3$$

compounds. Thus to isolate 46.7 actives requires

$$2 + 1000 + 20 \times 68.3 = 2368$$

units of testing. Thus the yield is

$$\frac{46.7}{2368} \times 1000 = 19.7.$$

If you want to play around a little more with this example see what the yields would be with cut-off points at 3.5 or at 5.

What's yours? (Chapter 12, page 222)

If you order malt whisky every time, in the long run it comes with soda $\frac{1}{3}$ of the time, and without soda on the remaining occasions. Thus your euphoria expectation is

$$\tfrac{2}{3} \times 21 + \tfrac{1}{3} \times 4 = 15.33,$$

while for blended whisky it's

$$\tfrac{2}{3} \times 15 + \tfrac{1}{3} \times 12 = 14,$$

so in the long run we keep up our euphoria best by ordering malt every time.

If soda only goes in $\frac{1}{5}$ of the orders the expectation on ordering malts becomes

$$\tfrac{4}{5} \times 21 + \tfrac{1}{5} \times 4 = 17.6$$

and for blended it is

$$\tfrac{4}{5} \times 15 + \tfrac{1}{5} \times 12 = 14.4,$$

so the moral is stick to malt (which most sensible Scots do anyway).

If things go the other way and the barmaid ruins too many

malt whiskies by putting in soda we may have to change our order. If, for example, she gets so absent-minded that she puts soda in 3 out of 4 the euphoria expectation for malt drops to

$$\tfrac{1}{4} \times 21 + \tfrac{3}{4} \times 4 = 8.25$$

while for blended it only drops to

$$\tfrac{1}{4} \times 15 + \tfrac{3}{4} \times 12 = 12.75.$$

It's time to sack the barmaid when she becomes this absent-minded.

More punishment

If you're really keen you might try a few more examples. They're not tied to any specific chapter this time, and there are no hints. If you can cope with these you've really got the feel of what statistics is about.

1: *Screwiness*. Mr Smith the ironmonger gets 70 per cent of his screws from Strongscrew Ltd and the remaining 30 per cent from Wobblescrew and Co. He mixes them together in one big box. McTavish buys one screw. What is the probability it was made by Wobblescrew? McTavish is very irate when he finds the screw is unslotted so he takes it back to Smith who informs him that only 1 per cent of the screws from Strongscrew are unslotted, but 20 per cent of the screws from Wobblescrew are. 'Therefore,' says Smith, 'it's pretty certain this is a Wobblescrew.' Thereupon McTavish, who fancies himself as an amateur statistician, does a quick calculation and announces that in his opinion the probability the screw was made by Wobblescrew is $\tfrac{60}{67}$. Is he justified in this claim? If not, what should the probability be?

2: *Political dilemma*. A political party leader tries to stop one of his followers making rash promises and keeps a careful record of how many he makes each day after he issues a warning. On day 1 he makes 4, on day 2 he makes 3, on day 3 he makes 3 and on day 4 he makes 2. The leader is delighted, and being an amateur statistician he denotes the number of rash promises by y and the day by x and calculates the regression of y on x as

$$y = -0.6x + 4.5.$$

Check if he has done his calculations correctly. Plot the points on graph paper and see if the result looks sensible. Would you agree that the leader has some grounds for optimism because of the 'negative' slope trend implied by the coefficient of x?

On day 5 the member makes 10 rash promises and the party leader amends his regression equation with the additional values $x = 5$, $y = 10$ and gets

$$y = 1.1x + 1.1.$$

Check if he is correct. If he is the equation implies an upward trend in his member's rashness. The party leader curses statistics for building up his hopes and then dashing them.

If you were the party leader's statistical adviser how would you explain to him that statistics is not the trouble, but that the way he is using his statistical ideas is not really very helpful?

3: *On the boss's knee.* One hundred businessmen, all with glamorous secretaries, are given two questions. They are told to answer the first if their birthday is in July, August or September and otherwise the second. The questions are

1. Does the month of your birthday have a letter r in it?
2. Have you ever had an affair with your secretary?

There are 37 'yes' answers. Would you agree that it was reasonable to assume from this data that about 40 per cent of these business men had had affairs with their secretary?

4: *Mastermind.* Muddlemind Calculators Inc. claim that after half an hour's training the average time taken by a statistics student to calculate the sum of the squares of ten four-digit numbers on one of their machines is less than 20 seconds. A random sample of 10 students from Professor Hyptest's class take the following times in seconds

19, 24, 25, 22, 24, 28, 26, 23, 18, 24

How do you feel about the manufacturer's claim in the light of this evidence?

5: *Side splitting*. A retailer of ladies underwear complains to a manufacturer that 10 per cent of the items he supplies are brought back by customers after one week with a complaint that they have split down the seams. The manufacturer promises the retailer he will do better and suggests that to prove that a large batch is acceptable the retailer should get the girls on his staff to wear a sample of five for a week. If none of the seams split in these five he convinces the retailer that the batch is worth taking. Do you think the retailer should be convinced? If 10 per cent of a large batch have faulty seams what is the probability that a test of a sample of five will show no faults?

Here's hoping that reading this book has helped you to get a bit more meaning out of figures.

Bibliography

AITCHISON, J., (1970), *Choice against Chance*, Addison-Wesley.

BARNARD, G. A., (1959), 'Control charts and stochastic processes', *Journal of the Royal Statistical Society*, B, vol. 21, no. 2, pp. 239–71.

BARTHOLOMEW, D. J., and BASSETT, E. E., (1971), *Let's Look at the Figures*, Penguin.

BLACKITH, R. E., and REYMENT, R. A., (1971), *Multivariate Morphometrics*, Academic Press.

CAMPBELL, R. C., (1967), *Statistics for Biologists*, Cambridge University Press.

CAMPBELL, C., and JOINER, B. L., (1973), 'How to get the answer without being sure you've asked the question', *American Statistician*, vol. 27, no. 5, pp. 229–31.

COCHRAN, W. G., (1963), *Sampling Techniques*, Wiley, New York.

COCHRAN, W. G., and COX, G. M., (1957), *Experimental Designs*, 2nd edition, Wiley, New York.

COX, D. R., (1958), *The Planning of Experiments*, Wiley, New York.

COX, C. P., (1968), 'Some observations on the teaching of statistical consulting', *Biometric*, vol. 24, no. 4, pp. 789–801.

DANIEL, C., and WOOD, F. S., (1971), *Fitting Equations to Data*, Wiley, New York.

DAVIES, O. L., and GOLDSMITH, P. L., (Eds.), (1972), *Statistical Methods in Research and Production*, 4th edition, Oliver & Boyd.

DRAPER, N. R., and SMITH, H., (1966), *Applied Regression Analysis*, Wiley, New York.

DUCKWORTH, W. E., (1968), *Statistical Techniques in Technological Research*, Methuen.

DUNNETT, C. S., (1972), in *Statistics: A Guide to the Unknown* (edited by J. Tanur), Holden Day, San Francisco.

Facts in Focus (1974), 2nd edition, Central Statistical Office/ Penguin Books.

FISHER, R. A., and YATES, F., (1963), *Statistical Tables for Biological Agricultural and Medical Research*, 6th edition, Oliver & Boyd.

GEFFEN, G., BRADSHAW, J. L., and NETTLETON, N. C., (1973), 'Attention and hemispheric differences in reaction time during simultaneous audio-visual tasks', *Quarterly Journal of Experimental Psychology*, vol. 25, no. 3, pp. 404–12.

HUFF, D., (1973), *How to Lie with Statistics*, Penguin.

KEMPTON, R. J., and MAW, G. A., (1974), 'Soil fumigation with methyl bromide', *Annals of Applied Biology*, vol. 76, no. 2, pp. 217–29.

LINDLEY, D. V., and MILLER, J. C. P., (1970), *Cambridge Elementary Statistical Tables*, Cambridge University Press.

MORONEY, M. J., (1965), *Facts from Figures*, Penguin.

SPRENT, P., (1969), *Models in Regression*, Methuen.

THURSTONE, L. L., (1947), *Multiple-Factor Analysis*, University of Chicago Press, Chicago.

WETHERILL, G. B., (1969), *Sampling Inspection and Quality Control*, Methuen.

WILLIAMS, C. B., (1970), *Style and Vocabulary*, Griffin.

WILLIAMS, E. J., (1959), *Regression Analysis*, Wiley, New York.

YATES, F., (1960), *Sampling Methods for Censuses and Surveys*, 3rd edition, Griffin.

Index